# LONDON LOCAL TRAINS
## IN THE 1950s AND 1960s

First published in Great Britain in 2016 by
PEN & SWORD TRANSPORT
an imprint of
Pen & Sword Books Ltd,
47 Church Street,
Barnsley,
South Yorkshire,
S70 2AS

Copyright © Kevin McCormack, 2016

A CIP record for this book is available from the British Library.

ISBN 978 1 47382 721 9

The right of McCormack, to be identified as the Author of this Work has been asserted by him in accordance with the Copyright, Designs and Patents Act 1988.

All rights reserved. No part of this book may be reproduced or transmitted in any form or by any means, electronic or mechanical including photocopying, recording or by any information storage and retrieval system, without permission from the Publisher in writing.

Printed and bound in India by Replika Press Pvt. Ltd.

Pen & Sword Books Ltd incorporates the Imprints of Pen & Sword Aviation, Pen & Sword Maritime, Pen & Sword Military, Wharncliffe Local History, Pen & Sword Select, Pen & Sword Military Classics and Leo Cooper.

For a complete list of Pen & Sword titles please contact
Pen & Sword Books Limited
47 Church Street, Barnsley, South Yorkshire, S70 2AS, England
E-mail: enquiries@pen-and-sword.co.uk
Website: www.pen-and-sword.co.uk

**Title page:** Three years before electrification of the former LTSR, Standard class 4 tank No 80100 leaves Barking station with a train from Fenchurch Street in winter 1957/8. In the background are the LT District Line tracks to Upminster. The locomotive now resides on the Bluebell Railway. *(Julian Thompson/Online Transport Archive)*

**Front cover:** The replacement of the local steam trains out of Paddington took the form of 3-car diesel multiple unit (DMU) sets built by the Pressed Steel Company in 1959-61 (subsequently BR class 117). Several of these DMUs have been preserved because they were highly successful and clocked up a remarkable working life of some forty years. This view at West Drayton depicts two 3-car sets headed by unit No 51349 in brand new condition. *(Marcus Eavis/Online Transport Archive)*

**Rear Cover (main picture):** Heading west in the evening sunshine, a train of ex-LNWR Oerlikon electric stock stops at Caledonian Road & Barnsbury on the former North London Railway on its journey from Broad Street to Bushey. *(Julian Thompson/Online Transport Archive)*

# LONDON LOCAL TRAINS
## IN THE 1950s AND 1960s

by Kevin McCormack

Boy seems to have problem with smuts in the eye as 2-6-2 tank No 41299 pulls out of Christ's Hospital with a Horsham - Guildford train in May 1965. *(Neil Davenport)*

# Introduction

This colour album of 1950's and 1960's images covers non-express trains working in and out of London termini, along with a selection of feeder services operating in roughly a 40 mile radius of the Capital. The trains featured are therefore semi-fast passenger, suburban passenger and freights. The advantage of casting the net beyond services in and out of London itself is to increase the variety of locomotive types and diesel/electric units featured, hopefully making the book more interesting to readers.

Conscious of the fact that much interest in this type of book is generated by railway modellers, I have endeavoured to use photographs with identifiable land marks, e.g. stations and signal boxes, wherever possible. Furthermore, every effort has been made to show material which has not been seen before in publications or on the Internet. However, it is perhaps worth mentioning that many railway photographers in the 1950s and 1960s travelled together so I cannot guarantee that occasionally an image very similar to one featured in this book has not been seen before. Also, since not all photographers were meticulous about their record of the trains they photographed, and in other cases preserved images taken by deceased photographers have been used with no information available, caption detail on individual workings is somewhat variable.

I make no apology for the fact that I am a steam enthusiast and therefore most of the pictures feature steam at work. I can raise some enthusiasm for vintage electrics, several of which are featured, but there is only a token representation of what I regard as more modern electrics and diesels, and definitely no unsightly yellow fronts!

The history of London services is a complicated one, not least because the railway companies were normally obliged for various reasons to site their London terminus initially on the outskirts of the City/West End until they were able to reach their preferred central locations. The first of London's current termini was London Bridge which was officially opened on 14 December 1836, from where services operated by the London and Greenwich Railway ran for some four miles to Deptford. Indeed, the first part of the line, just two miles of it, had opened in February 1836. London Bridge has however been rebuilt several times since 1836 and is therefore unrecognisable from its original state.

The first mainline London terminus was Euston which opened on 20 July 1837 and in the following year the London and Birmingham Railway began operating services between the two cities. The original station was controversially demolished in 1961/2, together with the seventy foot high entrance Doric Arch, with just the two lodges on Euston Road remaining. The present airline terminal style building is scheduled for replacement as part of the proposed High Speed 2 development which should also see the reconstruction of the Arch, over sixty per cent of the original stone cladding having been retrieved from the bottom of a waterway (the Prescott Channel) in Bow.

The next oldest London terminus is Fenchurch Street station. This opened on 20 July 1841 when the London and Blackwall Railway extended westwards from its previous terminus at Minories. The present station building dates from 1853/4, just being pipped in terms of longevity of original buildings still recognisable today by King's Cross. This station was opened by the Great Northern Railway on 14 October 1852, replacing the former terminus at Maiden Lane.

Jumping forward to 1863, the world's first underground railway opened on 10 January of that year, with the Metropolitan Railway operating services over a 3.75 mile route between Paddington (Bishop's Road) via Baker Street and King's Cross to Farringdon Street. Five years later, Baker Street became a terminus following the inauguration of services to Swiss Cottage on 13 April 1868. This was the first section of the line which eventually extended some fifty miles from Baker Street to Verney Junction and which spawned new suburbs in its wake, known as "Metroland".

To complete the picture, the remaining London termini opened as follows, albeit few with their original buildings/structures still intact today: Waterloo – 1848 (replacing Nine Elms, opened in 1838), Paddington – 1854 (replacing Bishops Road, opened in 1838), Victoria – 1860, Charing Cross – 1864, Cannon Street – 1866, St Pancras – 1868, Liverpool Street – 1874 (replacing Shoreditch, opened in 1841), Blackfriars (St Paul's) – 1886 and Marylebone – 1899 (replacing Canfield Place, opened in 1893). Finally, the opening dates for two London termini which closed in the latter part of the twentieth century were Broad Street – 1865 and Holborn Viaduct – 1874. Services were withdrawn from these stations on 30 June 1986 and 29 January 1990 respectively.

The population of London was already increasing considerably when the railway era began and the new lines serving the Capital encouraged urban development which started to swallow up the outlying villages and fields. Although the first railway entered London on a viaduct, most carved their way through residential areas, forcing inhabitants to move, usually further out of London. As a form of compensation, some railway companies were obliged to provide cheap, workmen's tickets.

The onset of electrification accelerated the growth of London "suburbia". Electric tube trains were introduced in 1890 and electric trams quickly followed. The latter had a particular impact on the railway companies operating in the south of London whose answer to this competition was to electrify their surface suburban services, starting in 1909.

Although the Great Western Railway was formed originally to provide a train service between London and Bristol, like other railway companies it opened in stages, with the initial section from Paddington - Maidenhead Bridge (Taplow) opening on 4 June 1838. As a consequence, the GWR can probably claim to have operated the earliest ground-breaking London local train service. This took place on 13 June 1842 when 23-year-old Queen Victoria travelled from Slough to Paddington (Bishop's Road, just west of Bishop's Bridge, the present station not opening until 29 May 1854). Her Majesty was quick to see the advantages of rail travel in terms of speed and comfort (of course, she was experiencing the benefits of Brunel's Broad Gauge) and is said to have been particularly enthralled by the view across the Brent Valley from the Wharncliffe Viaduct on the Southall/Hanwell border (see page 15). The travelling time for the 18-mile journey was a mere 25 minutes so it is no wonder that railways received immediate royal approval and gained instant respectability.

The majority of railway lines in and around London were built in Queen Victoria's reign, the last one to enter the capital (not counting HS1 in 2003) being the Great Central Railway (the original High Speed (HS) line built for carrying continental traffic) in 1899. A few new suburban lines were opened in the twentieth century, the final one being Motspur Park - Chessington in 1939. The onset of the Second World War followed by the Green Belt legislation prevented this route continuing to Leatherhead, which would have carved its way through Malden Rushett and Ashtead Woods, accompanied by the inevitable suburban housing.

The Beeching axe of the 1960s had a relatively limited impact on London local passenger traffic although several branch lines in the Home Counties did succumb. The biggest change however probably concerns freight traffic. The railways quickly replaced the canals as the main method of transporting goods in the nineteenth century and it was well into the twentieth century before road transport became the preferred means. Virtually every station had its own goods yard (the land now usually used for station car parks) and, as depicted in this book, many freight trains carried mixed goods in a variety of wagons and vans, often stopping at every local station on a daily pick-up basis, and sometimes ending up in large marshalling yards. This is a far cry from the present-day movement of goods by rail which is largely confined to bulk loads, often in large containers (and without the familiar guard's van attached at the rear).

Examples of the types of train featured in this album, so familiar in the 1950s and 1960s, can only be found now on the preserved heritage railways, but fortunately there are many of these around Great Britain, including a few in the Home Counties. The transition to a modern railway system, combined with de-nationalisation, has been dramatic and I hope that fond memories of the railways of our childhood and youth in and around London will be pleasantly awakened by the images in this book.

I would like to record my thanks to the many people, whether individuals or organisations, who have provided photographs: Alan Sainty, Neil Davenport, Bill Piggott, Jim Oatway, Gordon Wells, the Light Rail Transit Association (LRTA) and the main source, the Online Transport Archive, a charity which preserves motion and still photographic collections. Thanks are also due to Bob Bridger for allowing me access to Charles Firminger's material and to Mike Eyre for revitalising some images.

Finally, I should mention that the images in this book are arranged in regional order, Western, Southern, London Midland and Eastern, and ends with London Transport. The regions are divided into segments relating to various divisions or different lines, as appropriate, which should provide a degree of logic to the batching. Also, locations are generally in order outward of London and are listed alphabetically on pages 167/8.

*Kevin R McCormack, Ashtead, Surrey*
*May 2014*

Before we embark on our BR region tour around London and the Home Counties, here is a shot of a lucky locomotive which escaped the scrapman's torch. Former LSWR M7 0-4-4 tank No 30053 was built at Nine Elms in 1905 and remained in stock until May 1964. Three years later it was shipped to Steamtown, Vermont in the USA, returning to our shores in 1987 and going to its present home on the Swanage Railway. From January to September 1963, the engine was based at Tunbridge Wells West and the view depicts it on a "motor train" (push and pull) service from Oxted entering Ashurst (Kent). The station remains open today but the original buildings, dating from the introduction of services by the LBSCR in 1888, were demolished in 1983 and replaced by "bus shelters" on the platforms. The railway from Oxted through Ashurst originally went to Lewes but this was cut back to Uckfield from 6 January 1969 and the line from Eridge (the next station south of Ashurst) to Tunbridge Wells West, where this train is destined for, was closed on 8 July 1985 but has now reopened under the auspices of the heritage Spa Valley Railway. *(Phil Tatt/Online Transport Archive)*

# Abbreviations

Several abbreviations have been used in the text, all of which will be recognised by the majority of readers, but for completeness they are as follows:

| | | | |
|---|---|---|---|
| AEC | Associated Electrical Company | LMR | BR London Midland Region |
| BIL | Bi-lavatory unit | LMS | London, Midland & Scottish Railway |
| BR | British Railways | LNER | London & North Eastern Railway |
| DMU | diesel multiple unit | LNWR | London & North Western Railway |
| EMU | electric multiple unit | LSWR | London & South Western Railway |
| EPB | electric-pneumatic brake | LT | London Transport |
| ER | BR Eastern Region | LTSR | London, Tilbury & Southend Railway |
| GC | Great Central Railway/Great Central line | Met | Metropolitan Railway/Metropolitan Line |
| GER | Great Eastern Railway | NLR | North London Railway |
| GNR | Great Northern Railway | SECR | South Eastern & Chatham Railway |
| GWR | Great Western Railway | SER | South Eastern Railway |
| 2 HAL | half-lavatory | SR | Southern Railway/ BR Southern Region |
| 2 HAP | half-lavatory with electro-pneumatic brake | SUB | Suburban |
| LBSCR | London, Brighton & South Coast Railway | WCML | West Coast Main Line |
| LCDR | London Chatham & Dover Railway | WR | BR Western Region |

Starting with the WR, empty stock movements between Old Oak Common carriage sidings and Paddington were usually worked by various types of pannier tank, such as Collett condensing tank No 9700. This is seen entering Paddington station on 15 April 1961, flanked by two distant Hawksworth heavy pannier tanks. No 9700 was designed to take freight trains to the GWR's goods depot beneath Smithfield (meat) Market, some three miles east of Paddington. Access was via the Met's Underground lines, requiring GWR locomotives to be fitted with condensing apparatus for working through the tunnels. By the early 1930s the existing locomotives used on these freights were wearing out and No 9700 became the guinea pig for a replacement type. Built by Beyer Peacock, this engine entered service as No 8700 in February 1931. In March 1932 it was converted into a condensing locomotive, spawning a further ten such pannier tanks, Nos 9701-9710, constructed at Swindon in 1933, and No 8700 was then renumbered in January 1934. *(Alan Sainty collection)*

One of the WR's Blue Pullman diesel sets passes stock hauled by Modified Hall 4-6-0 No 7900 *Saint Peter's Hall* near Royal Oak, on the outskirts of Paddington, in 1961. By this time, there were three blue Pullman services, the Birmingham Pullman, the Bristol Pullman and the South Wales Pullman, the first two trains having been introduced on 12 September 1960. The Pullmans were certainly visually attractive but were not very reliable, being underpowered and with a tendency towards rough riding. They were withdrawn by 1973 but left behind an important legacy, influencing the design of the highly successful Inter City 125s which are still operating today. As regards the role being performed here by No 7900, this is unclear as it is carrying a headlamp in the position for shunting, an unlikely activity in this particular location. (*Harry Luff/Online Transport Archive*)

Old Oak Common was the depot serving Paddington but, in addition to having an allocation of express locomotives, it also housed more menial classes for general duties. One such engine, pannier tank No 3620, stands alongside the shed in January 1965 at the head of a train of oil tank wagons, evidence of the advancing dieselisation programme. The locomotive is in typically unkempt condition for this final year of WR steam, its number chalked on in several places following removal of its cabside number plate. Unauthorised access to the depot in steam days was often achieved by finding a hole in the fence along the towpath of the nearby canal! (*Author*)

Another unkempt locomotive typifying the decline of WR steam, 2-8-0 No 3851 brings a freight train off the Greenford Loop and towards West Ealing Station (seen in the background) in March 1965. The picture was taken from "Jacob's Ladder", the footbridge where the author spent much of his misspent youth trainspotting! The 2884 class to which No 3851 belonged was Collett's development of Churchward's earlier 28XX class (Nos 2800-2883), designed for heavy freight work. Several examples of both types are preserved, all but one having ended up in Dai Woodham's rest home for redundant engines (aka Barry Scrapyard) and subsequently extracted. (*Author*)

Playing Cowboys and Indians in the fields between Castlebar Park Halt and Drayton Green Halt on the Greenford branch in the 1950s was a regular pastime for the author, interrupted by the steam push and pull passing by, hauled by a diminutive 14XX 0-4-2 tank or a more powerful 54XX pannier tank. However, pleasure turned to pain on 25 August 1958 when diesels took over this Ealing Broadway - Greenford service (now Paddington - Greenford). The line had opened in 1903 to enable a circular service to and from Paddington to be operated for the Royal Agricultural Show located on land at Twyford Abbey (the district subsequently being called Park Royal). Normal branch services started in 1904. Shortly before the end of steam operation, No 1426 waits at Drayton Green Halt before propelling its auto trailer (W220 *Thrush*) around the east curve to West Ealing station. Note the red cabside numberplate, a short-lived fad. (*Julian Thompson/Online Transport Archive*)

Until DMUs started to displace steam hauled compartment stock in the late 1950s, most of the WR suburban trains into and out of Paddington were hauled by the powerful 61XX 2-6-2 tanks which replaced the similar looking County 4-4-2 tanks in the early 1930s. This winter 1957/8 view looking north depicts a green-liveried "tanner-oner" (as we loco spotters called the class) on an up local to Paddington crossing Brunel's splendid Wharncliffe Viaduct over the Brent Valley at Hanwell. Dating from 1836/7, the viaduct was Brunel's first major structural design and was named after the chairman of the parliamentary committee that steered the GWR bill through Parliament and whose coat of arms are visible on the brickwork below the engine's bunker. Although the viaduct now carries overhead electrification support columns, these have been sympathetically sited so that they are scarcely any more obtrusive than the previous telegraph poles. (*Julian Thompson/Online Transport Archive*)

In the final year of WR steam, LMR locomotives became increasingly evident in the London area on freight work. Looking comparatively smart for the time (August 1965) Stanier 8F No 48012, dating from 1936, is approaching Southall engine shed with a westbound goods. Sir Willliam A Stanier started his railway service with the GWR before moving to the LMS and it is hardly surprising therefore that his designs have a look of the GWR about them. On the other hand, Hawksworth's County 4-6-0s were fitted with boilers which had much in common with the Stanier 8Fs, Hawksworth seemingly taking more than a passing interest in the batch of 8Fs being built by the GWR at Swindon Works during the Second World War. (*Author*)

Viewed from the footbridge where the seeds of the Great Western Society were first sown by four local schoolboys, Castle class 4-6-0 No 5023 *Brecon Castle* brings Whitland-bound milk empties past Southall shed on 10 July 1962. Milk traffic to London from the West Country and South Wales provided the WR with lucrative business and express locomotives of the Castle or King classes were normally provided for what were regarded by the GWR and WR as prestigious services. No 5023 was built in 1934 and withdrawn in February 1963, some six months after this photograph was taken. The locomotive was scrapped but eight members of the 171-strong class have been preserved. (*Alan Sainty collection*)

Looking westwards from Southall footbridge, 2-6-2 tank No 6117 stands at Southall station with a Paddington-bound local in 1959. The railway line from Paddington to Maidenhead Bridge (Taplow) opened in 1838 but it was almost another year before Southall station was built. Much of the background behind the train remains today: the right hand portion of the station building, the gas holder (dating from 1932) and the castellated brick water tower built in 1903 and now converted into apartments. However, this scene can be expected to alter significantly as a result of Crossrail which will burrow under central London, linking Maidenhead (or possibly even Reading) and Heathrow in the west to Shenfield and Abbey Wood in the east. *(Julian Thompson/Online Transport Archive)*

Recently arrived from the West Country following dieselisation, light Prairie tank No 5531 was one of a small batch of these locomotives allocated to Southall depot in 1964, bringing some welcome variety to the somewhat uninspiring stud of 61XX large Prairie tanks and pannier tanks which epitomised this largely freight shed. In this view dating from December 1964, No 5531 is about to haul a freight train from West Drayton sidings adjacent to the Coal Depot opened in 1963 which its owners, the National Coal Board, claimed to be the largest mechanical depot for handling household coal in Europe. (*Author*)

Ex-GWR prairie tank No 6147 departs from Cowley with a stopping train to Paddington from Uxbridge Vine Street in summer 1957. The normal WR 5-carriage suburban set has been strengthened with an additional coach behind the engine. Until 1939, the town of Uxbridge boasted three railway stations serving different lines, the first to close being the branch from the Birmingham main line to Uxbridge High Street which was meant to join up with the Uxbridge Vine Street/West Drayton branch but the connection building works were never completed. Neither line was able to compete successfully with the Metropolitan and Piccadilly line electric services to London from the more centrally sited Underground station. (*Julian Thompson/Online Transport Archive*)

West Drayton was the junction, not just for the Uxbridge Vine Street branch but also one to Staines West, the latter closing to passengers in 1965. However, whereas part of the Staines West branch (as far as Colnbrook) remains open for freight, the line to Uxbridge has been totally abandoned following cessation of parcels traffic from 13 July 1964 (passenger services were withdrawn from 10 September 1962). In this view, ex-GWR diesel railcar No W31W from Southall shed is seen leaving Uxbridge Vine Street station on the 7 minute, 2 mile journey to West Drayton, calling at the only intermediate station, Cowley, on its way. (Julian Thompson/Online Transport Archive)

Hawksworth's 2-cylinder County class 4-6-0s were something of an enigma because there is some doubt over why they were built. Classified as 6MT (ie powerful mixed traffic), the thirty members of the class were constructed between 1945 and 1947 alongside new Castles and Modified Halls which were more than capable of covering the passenger and freight duties assigned to the Counties. Their most distinctive visual feature was the combined splasher over the driving wheels with the resultant straight nameplate, Hawksworth presumably obtaining the idea from the semi-streamlined Castle and King (Nos 5005 and 6014) of the 1930s. This view depicts No 1020 *County of Monmouth* at Slough on 27 September 1962. (*Alan Sainty collection*)

The GWR's diesel railcar fleet of 38 vehicles included two express parcels cars. These were No 17, the last of the pre-war streamlined batch, and angular-shaped No 34. This latter vehicle was based at Southall and was hated by Ealing trainspotters like myself because of its irritatingly frequent appearances (and for being a diesel, albeit an elderly one!). Built at Swindon Works in 1941 and fitted with two AEC bus engines, W34W, as it came to be numbered, was withdrawn in September 1960 and is pictured here at Slough on 14 September 1959 in rather jaded condition. (Ian Dunnet/Online Transport Archive)

One WR branch in the London/Home Counties area which has survived to the present day is that from Maidenhead to Bourne End and Marlow. Until 7 July 1962, the "Marlow Donkey" (as it was affectionately known) was a steam push and pull (auto train) in the hands of a 14XX 0-4-2 tank from Slough shed, succumbing fairly late in the day to replacement by DMU. This could have been due to the volume of freight traffic at Marlow, by then on the decrease, which often necessitated the "Fourteener" shunting and assembling a mixed train. In summer 1961, No 1421, coupled to its auto trailer and a van, was photographed by the 14-year-old author taking water at Maidenhead in unusual circumstances, the details of which have been forgotten with the passage of time. Clearly, the van could not have been propelled from Marlow/Bourne End to Maidenhead, being non-compliant with auto working, and the same problem would have arisen if the engine were intending to make a journey from Bourne End to Marlow (unless it ran round and pulled the van). *(Author)*

A visit to Reading on 3 January 1965 to witness the last day of scheduled steam services on the Redhill - Reading (Southern) line resulted in the author straying onto the nearby platforms of Reading (General) station. There, a particularly shabby Modified Hall class loco, No 6986 *Rydal Hall*, but still carrying its number plates and name plates, was standing ready to haul some rolling stock out of the adjacent sidings. Reading (General), the suffix being added in 1949 to distinguish it from the Southern station, was opened by the GWR in 1840, and fifteen years later the SER opened their station as the terminus of their route to Guildford and Redhill. Later still, the LSWR also used it for their services from Waterloo. Both these services were diverted into Reading (General) station from 6 September 1965, whereupon Reading South (later renamed Reading (Southern)) was closed to passengers. (*Author*)

In the period covered by this book there were two main lines served by trains to and from Paddington, one to the west, already featured in the preceding photographs, and the other to the north-west, the lines bifurcating at Old Oak Common. This latter line now only has one train a day in each direction, just for route familiarisation purposes. This photograph, taken on 2 June 1965, shows the celebrated Mon-Fri 4.15 pm Paddington - Banbury train, the last scheduled steam-operated service from Paddington, running alongside the LT Central Line tracks near North Acton headed by No 6903 *Belmont Hall*. The locomotive is in typically deplorable external condition for the period, with no name plates or number plates, and the steam service has only seven more days of operation. Friday, 11 July marked the last day and photographs appeared in the national press and on television, such was the media interest in this modest but ultimately historic service. (*Author*)

An Ealing Broadway - Greenford auto train service, with No 1474 in charge, stands in the bay platform at Greenford in 1958, shortly before dieselisation. This modernisation saw Southall depot lose all its 14XX 0-4-2 tanks except No 1474, which was the last of the class to be built. Southall retained this locomotive for several years and loaned it to Slough shed for a few weeks in 1961 for use on the *Marlow Donkey*. No 1474 was eventually withdrawn in September 1964 from Gloucester shed, its last duties being on the *Chalford Flyer*. The bay platform at Greenford stands between the LT Central Line tracks to West Ruislip. The WR Birmingham main line is on the extreme left. (*Geoff Morant/Online Transport Archive*)

Here are two more pictures of the 4.15 pm Paddington - Banbury (shown on the Departure Board at Paddington as terminating at King's Sutton to discourage passengers for Banbury from taking it since the 4.10 pm provided a much faster journey and even the 5.10pm reached Banbury before the 4.15pm!). On 17 September 1964, scruffy Castle class 4-6-0 No 7024 *Powis Castle* was photographed from Carr Road footbridge, east of Northolt station. In contrast, No 7029 *Clun Castle* was a joy to behold on 19 May 1965 at Beaconsfield, having been spruced up for the benefit of Peter Lemar (a Great Western Society dignitary) who was travelling on the footplate and has stepped down to take a picture. *(Bill Piggott; Author)*

Hawksworth Modified Hall No 6990 *Witherslack Hall* from Old Oak Common shed brings an up freight through Beaconsfield (the station is in the background) on 13 September 1962. Entering service in April 1948, the engine was chosen by the WR to take part in the 1948 Locomotive Exchanges in the mixed traffic designs category. No 6990 was trialled on the former GC main line and so it is very appropriate that, following its withdrawal in December 1965 and a sojourn at Barry scrapyard, it should be preserved by the heritage Great Central Railway. Beaconsfield station, on the joint GW/GC line, was opened in 1906 (see opposite page). The two centre through lines were removed in 1974. *(Alan Sainty collection)*

High Wycombe was originally merely a station on the line from Maidenhead which was subsequently extended to Princes Risborough and, via Thame, to Oxford, thereby providing the GWR with a shorter route to Oxford than the route via Reading and Didcot. However, High Wycombe's importance increased very significantly following the opening of the Great Western & Great Central Joint Railway from Northolt Junction to Ashendon Junction in 1906, bringing GC trains from Marylebone and, by 1910, GWR trains from Paddington via North Acton and Greenford. This photograph depicts large prairie tank No 6117 leaving High Wycombe for Aylesbury around 1960, having arrived from Maidenhead and Bourne End. (*Author's collection*)

Freight services were withdrawn between Bourne End and High Wycombe on 18 July 1966 (nearly four years before the cessation of passenger services) and yet only some seven months earlier Wooburn Green goods yard is a hive of activity. The date is 28 December 1965, three days before the official end of WR steam, and Stanier Black 5 No 44860, with a Coventry shedcode (2D), has just arrived with a freight from High Wycombe while immaculate large prairie tank No 6106 (earmarked for purchase by a Great Western Society member) prepares to leave for Bourne End and Maidenhead with a freight. Currently, there is pressure to reinstate the abandoned line. (*Author – both*)

Princes Risborough, on the WR's Birmingham main line, was once a busy junction, with branches, since closed, to Thame/Oxford and to Chinnor/Watlington (part of which is now preserved by the Chinnor & Princes Risborough Railway), and one which is still open – to Aylesbury. Auto trains operated many of the services, some of which continued beyond Princes Risborough, e.g. to High Wycombe, and there were also some through trains emanating from the West of England mainline via Maidenhead. These views depict 0-4-2 tank No 1440 and auto trailer W221 *Wren* in 1962, about to depart for Aylesbury, and 2251 class 0-6-0 No 2270 bringing in a pick-up freight from the Banbury direction in 1959. (*Marcus Eavis/Online Transport Archive; Harry Luff/Online Transport Archive*)

Moving on to the SR and beginning with the South Western division (ex-LSWR), we start with a reminder of the old manually-operated wooden Departure Board at Waterloo station which used to stand between Platforms 6 and 7. The displays were changed by staff operating levers at low level, one of which can be clearly discerned through the gap in the railings. Waterloo was opened as Waterloo Bridge station by the LSWR in 1848 after the line was extended over arches from Nine Elms. However, the LSWR intended it to be a through station for a line to the City of London and enlargements were added piecemeal until the LSWR's Waterloo & City underground railway to the Bank was opened in 1898, thereby meeting the LSWR's goal of reaching the City. Accepting then that Waterloo was always to be a terminus, a new station was built, coming into use between 1910 and 1922. (Neil Davenport)

Southern electrification was inaugurated by the LBSCR in 1909 using the overhead AC system whereas the LSWR adopted the third rail DC system when it introduced electrified services in 1915. Upon grouping in 1923, the new Southern Railway adopted the LSWR system and extended this to former SECR lines as well as converting ex-LBSCR routes. Much of the early SR third rail electric stock was built using modified wooden bodied steam stock mounted on new underframes. These EMUs were formed into 3-car motor units and 2-car trailer sets, but eventually 4-car sets became the standard for suburban services. The last of the 4-SUB electric units created by converting former LBSCR steam stock in the 1920s were withdrawn between 1958 and 1960. This one, No 4517, was photographed at Waterloo on 13 September 1958 on a Hampton Court service. (*Marcus Eavis/Online Transport Archive*)

Brent sidings at Cricklewood on the former Midland Railway main line were regularly visited by SR locomotives hauling cross-regional transfer freights. In this view dating from 1964, Bulleid Q1 0-6-0 No 33026 is preparing to head down the Dudding Hill loop line to Acton Wells Junction and onwards along part of the North London line to reach the SR. The locomotive belongs to the class of forty wartime Austerity 0-6-0s designed as powerful, lightweight engines for working the increasing number of vital freight trains operated by the SR during the Second World War. The locomotives were devoid of frills and could even be cleaned by passing through carriage washing plants. They were the most powerful 0-6-0s in Britain and all lasted into the 1960s. The doyen of the class is preserved as part of the National Collection. *(Alan Sainty collection)*

This SR-bound transfer freight has presumably travelled from Brent Sidings, Cricklewood, and could be heading for Norwood Yard. Motive power consists of C2X 0-6-0 No 32438 piloting an unidentified Q1 0-6-0. The C2X class comprised Marsh rebuilds of Billington's LBSCR C 2 class with larger boiler and extended smokebox. The train is passing through South Acton station on the former North London Railway Richmond branch, now marketed as the Overground. The station is still open but the adjacent former District Railway station behind the photographer was abandoned following the withdrawal of the single car shuttle service to Acton Town on 28 February 1959. (*Alan Sainty collection*)

Photographed from different directions by the signalman at West London Junction box which straddled the lines between Queenstown Road (formerly Queen Road) and Clapham Junction, class M7 0-4-4 tank No 30320 **(above)** brings empty stock from Waterloo to Clapham Junction carriage sidings on 5 June 1961 and class S15 4-6-0 No 30496 **(right)** hauls a freight from Feltham yards, destined for Nine Elms, on 10 July 1961. The overhead signal box opened on 11 February 1912, replacing a trackside box. London's skyline has changed dramatically since these pictures were taken with no sign of Canary Wharf, the Gherkin, the Shard or any other towering edifices. *(Jim Oatway – both)*

Although M7 tanks had been deployed on empty stock workings between Clapham Junction and Waterloo over many years, other steam tank locomotives were also used from time to time such as E4 0-6-2s, one of which, No 32497, is depicted in this view **(left)** at Clapham Junction on 18 March 1957. However, at times express engines suffered the indignity of reversing back to Clapham Junction hauling empty stock, as demonstrated **(above)** by this filthy looking unrebuilt "West Country" pacific, No 34023 *Blackmore Vale*, in 1966. However, luck was to shine on this engine for it now lives on the Bluebell Railway. (*Geoff Morant/Online Transport Archive; Marcus Eavis/Online Transport Archive*)

Standard class 4 tank, No 80113, brings a down van train into Clapham Junction station on 30 April 1966. It has just passed under A Box (originally East Box) which, on 10 May 1965, nearly ended up on the railway tracks. One of the girders supporting the gantry on which the signal box was positioned collapsed due to corrosion. As a result, the gantry had to be shored up, and to reduce the weight on the remaining supports the protective steel roof, fitted over the box as an air raid precaution during the Second World War, was removed, leaving just the framework visible in this picture. The Box was closed in 1990 and demolished. (*Marcus Eavis/Online Transport Archive*)

A Waterloo circular service via Richmond formed of an augmented 4-SUB unit, No 4506, utilizing bodywork from pre-grouping steam stock, enters Clapham Junction. The train is flanked by sets of post-war 4-SUBs built between 1946 and 1951 which were the mainstay of SR suburban electric services throughout the 1950s, 1960s and 1970s. In the centre of the picture at the opposite end of the signal gantry is the top of B Box dating from 1952 which replaced an earlier structure. (*Harry Luff/Online Transport Archive*)

Designed primarily to haul transfer freights across London, the powerful and fast 2-6-4 tanks of the W class, which numbered fifteen engines, were built at either Eastleigh or Ashford between 1932 and 1936. The engines were similar in appearance to the ill-fated River (K class) tanks which were rebuilt as 2-6-0 tender engines following the Sevenoaks crash in 1927 (see page 83) and are reputed to have incorporated some redundant K class parts. They were specifically non-passenger locomotives because, following some trials on passenger work, they were found to be unstable at high speed, like the previous K class. This photograph taken in 1957 from the north end of Clapham Junction station depicts a transfer freight hauled by No 31917 approaching Platform 17 from behind Clapham B signal box (out of view on the left). *(Alan Sainty collection)*

The last scheduled steam-hauled passenger service in the London area was the Clapham Junction - Kensington (Olympia) service which ran for the last time on 7 July 1967 before being dieselised. The service was not at that time advertised in the public timetables and was provided for Post Office Savings Bank staff, although anyone could travel on it if they got to hear of it. The trains only operated in the rush hour, making a one direction eight minute journey (i.e. it returned empty). This view shows the service, known colloquially as the "Kenny Belle", on arrival at Clapham Junction's Platform 17, with motive power provided by Standard Class 3 2-6-2 tank No 82019. *(Harry Luff/Online Transport Archive)*

Viewed from Alt Grove footbridge, an electric unit with three carriages formed from ex-LSWR steam stock augmented by a steel post-war 4-SUB trailer leaves Wimbledon station with a down service in 1957. On the left, part of Wimbledon B signal box (known as D box until March 1928) is visible. Interestingly, this box was manhandled 12 feet to the right in February 1928 to make room for two additional tracks to serve the Wimbledon - Sutton line then under construction. In the background is a rake of Italian vans which have arrived by ferry. They are standing in the former milk dock which was built in 1926 on the site of the "Volunteers Platform", previously used by the military and also for loading/unloading livestock. (*Julian Thompson/Online Transport Archive*)

4-SUB unit No 4381, dating from 1948, leaves Richmond station in summer 1955 on a down Waterloo circular service via Kingston and Teddington. The first ten 4-SUBs (4-coach suburban sets) were built between 1941 and 1945 and had domed cab roofs. Production of the vertical cab roof type seen here began in 1946 and continued until 1951. The RT bus on the road bridge is a 265 service on its way from Chessington (Copt Gilders Estate) to East Acton. Richmond's surface station building, visible in the background towards the left, was built in art deco style in 1937 and is largely unchanged today. The first station at Richmond opened in 1846, with services to Nine Elms (extended to Waterloo two years later). *(Frank Hunt/LRTA)*

Class N 2-6-0 No 31813 stands at Surbiton with a down train of empty wagons on 18 February 1962. Eighty of these Maunsell mixed traffic locomotives were built between 1917 and 1934, the first twelve being built under the auspices of the SECR including this example constructed in 1920 at Ashford Works. The station buildings are Grade II Listed and date from 1937 being another fine example of art deco style. It replaced an earlier structure, the rebuilding being preparatory to imminent electrification to Woking. *(Alan Sainty collection)*

It is 7 July 1967, just two days before the end of scheduled London steam, and Bulleid Light Pacific No 34052, travelling on the down main at Weybridge, is in superb external condition, despite having lost its *Lord Dowding* nameplates and being reduced to hauling a van train. On the extreme right is the corner of Weybridge (formerly Weybridge Junction) signal box which was closed on 22 March 1970, and which was situated virtually opposite the point where the Chertsey/Staines loop line diverged from the main line. The building behind the train is Weybridge's modest goods shed. *(Bill Piggott)*

Standard Class 5 4-6-0 No 73114 *Etarre* storms through West Byfleet in October 1965 with a van train. Twenty of the SR's Standard class 5s were given names from withdrawn King Arthur class 4-6-0s, 73114's name having previously been carried by Urie LSWR class N15 No 30751 built in 1922 and withdrawn in 1957. The passing train in this picture clearly holds no interest to the studious schoolgirl who is probably doing her homework (unlike today's schoolgirl who would likely be texting or playing an electronic game!). The contents of the station car park would now be much sought after by classic car enthusiasts. (*Alan Sainty collection*)

We have now reached Guildford, the extent of our journey from London on the SR South Western division, and see the 6.05pm departure to Horsham on 5 June 1965 hauled by LMS Ivatt class 2 2-6-2 tank No 41294. This locomotive was built by BR at Crewe in 1951 and belonged to a batch of thirty (Nos 41290-41319) which were allocated from new to the SR. In all, 130 were built by the LMS and BR between 1946 and 1952. The buildings behind the train were replaced in 1987 but the station footbridge in the background remains and is a public right of way, for which bridge passes are available to non-passengers. (Marcus Eavis/Online Transport Archive)

Moving over to the Southern Region's Central Division (ex-LBSCR) we see, **above,** class E4 0-6-2 tank No 32474 leaving London Bridge station with the 1.05 pm newspaper train on 26 April 1963. Built at the LBSCR's Brighton Works in 1898, the locomotive no longer survives but its numerical predecessor, No 32473, was saved and can be found on the Bluebell Railway in Sussex, sometimes bearing the name *Birch Grove,* depending on which livery it is carrying. More non-passenger activity at London Bridge is provided, **right,** by SECR class C 0-6-0 No 31690 with a train of ferry wagons, possibly originating from Southwark goods depot. (*Paul de Beer/Online Transport Archive; Geoff Morant/Online Transport Archive*)

2-HAL (two-car half-lavatory set, ie only one lavatory for the two units) No 2615 and 4-EPB (four-car electric-pneumatic brake) No 5017 stand at London Bridge High Level platforms in February 1958 beneath a footbridge which was about to be removed, this being the reason for the photograph being taken. The 2-HAL unit dates from 1938-9 and some of these sets lasted in service until 1971. The 4-EPBs were built between 1951 and 1957 and all were withdrawn by 1995. The High Level platforms served the SECR lines, as evidenced by EPB No 5017 which is working a Cannon Street - Gravesend or Maidstone West duty. This image is included in the Central Division section of the book for convenience. (Phil Tatt/Online Transport Archive)

Against a remarkably unattractive backdrop, two-car EPB unit No 5790 arrives at Battersea Park on its way from Victoria to London Bridge via Peckham Rye. This shuttle service, known as the South London Line, had operated in its entirety since 1867 until it ceased on 8 December 2012, partly replaced by less direct London Overground services. No 5790 was one of fifteen sets built at Eastleigh in 1954/5 for the Tyneside Electric lines and worked between Newcastle Central and South Shields until replaced by diesels in 1963. The units were then transferred to the SR and, after some necessary modifications had been made, entered service, as seen in this 1963 view. These ex-Tyneside 2-car EPBs were withdrawn in 1985. (*Marcus Eavis/Online Transport Archive*)

Brand new diesel electric multiple unit (DEMU) No 1313 stands at East Croydon in summer 1962. There were 19 of these 3-car units built, primarily for use on the Victoria - Oxted line, being able to operate over non-electrified as well as electrified sections. They also had the advantage of sharing some of the features of SR EMUs which was beneficial in terms of repair and maintenance. In this picture, the unit is heading away from the camera and the power car is at the front of the train (ie in the distance). When the unit was withdrawn in March 1994, this car was converted into tractor unit No 951070 and survives in preservation. Also preserved is a complete 3-car unit, No 1317, which has returned to its original haunts courtesy of the Spa Valley Railway. (Frank Hunt/LRTA)

Standard class 4 2-6-4 tank No 80065 takes the "reversible" line (i.e. used in different directions during the morning and evening peaks) out of East Croydon on 26 April 1962. The train is the 4.48 pm Victoria - Brighton stopping train composed of Maunsell stock including the bogie utility van. The cleaners have made a good job of the tank and bunker sides of the engine but appear to have given up on dealing with the rest of the locomotive which looks decidedly unkempt. The last three carriages of the train do not look too great either. The first station in Croydon opened in 1839 and was later renamed West Croydon to distinguish it from the second Croydon station (this one) opened in 1841 which, for the same reason, became East Croydon. The station building just visible behind the RT buses on the road bridge dates from the rebuilding in 1894/5. A steel and glass edifice replaced this structure in 1992. (*Charles Firminger*)

Oxted was originally an extremely busy interchange station served by trains operating between Victoria/ London Bridge and the South Coast. These through routes were victims of the Beeching Axe but ironically two preserved railway lines now operate over part of the abandoned lines, the Spa Valley Railway and the Bluebell Railway, while the national network south of Oxted is now confined to two truncated branch lines: to East Grinstead and to Uckfield. The view **above** shows Standard class 4 tank No 80011 leaving Oxted for Lewes via East Grinstead and Horsted Keynes on 9 May 1954, passing the ex-LBSCR timber signal box built in 1896 and closed in 1987. Just south of Oxted is Hurst Green Junction where the Uckfield line via Eridge (interchange for the Spa Valley Railway to Tunbridge Wells West) and East Grinstead line (interchange for the Bluebell Railway to Horsted Keynes and Sheffield Park) bifurcate.

The photograph, **right**, depicts No 80018 on 4 June 1962 hauling the 5.25 pm train from East Grinstead to Victoria, passing an ex-LBSCR lower quadrant signal at Hurst Green. The line to Eridge and onwards to Uckfield curves round to the left. (*Neil Davenport; Charles Firminger*)

A family runs to catch the Oxted push and pull, hauled by SECR class H 0-4-4 tank No 31518, before it sets off from Tunbridge Wells West in summer 1962. The building on the left, dating from 1886, is the only former LBSCR engine shed which is still fulfilling its original purpose, being now occupied by the Spa Valley Railway. It was previously a sub-shed of Brighton, classified 75F. The line from Eridge, on the Uckfield branch, to Tunbridge Wells West was closed to passengers on 6 July 1985. Re-opening as a heritage line began in 1997 and the entire section from Tunbridge Wells West to Eridge came into use in 2011. The impressive station building at Tunbridge Wells West is now a pub/restaurant. (*Marcus Eavis/Online Transport Archive*)

A train bound for Tunbridge Wells Central and Tonbridge on the Charing Cross - Hastings line, departs from Tunbridge Wells West behind SECR class L 4-4-0 No 31781. The photographer is standing on the bridge carrying Montacute Road over the railway and the train is about to enter the single track section through Grove tunnel and on to Grove Junction to join up with the mainline. This link line was imposed on the LBSCR by the rival SER which otherwise had threatened to build its own line to South Coast destinations served by the LBSCR. The carriages being hauled by No 31781 are Maunsell flat-sided stock designed for the narrow tunnels on the Tonbridge - Hastings line, the width restriction resulting from the tunnel builder cheating the SER by reducing the layers of bricks inside the tunnels which required more to be added following the collapse of Wadhurst Tunnel in 1862. The train pictured would have passed through one of the infamous tunnels (Somerhill) to reach Tonbridge. (Julian Thompson/Online Transport Archive)

Tunbridge Wells West was reached from several directions, one of which was by an eastwards extension of the Three Bridges to East Grinstead (high level) branch through Forest Row to Groombridge (now an intermediate station on the Spa Valley Railway). The entire line was closed on 1 January 1967, leaving at that date only the Groombridge to Tunbridge Wells West section open. On 5 September 1963 H class tank No 31005 was photographed approaching Cook's Pond viaduct over Wilderness Lake, between Dormans and East Grinstead, with the 6.13 pm from Oxted to Forest Row. The line from Oxted now no longer extends beyond East Grinstead. (*Charles Firminger*)

East Grinstead (low level) was also connected to a line to Lewes via Horsted Keynes, where there was a junction with a branch to Haywards Heath via Ardingly. The Lewes line closed on 17 March 1958, but an initial section was re-opened by the Bluebell Railway as early as 7 August 1960. The line to Haywards Heath, on which an electric train service between Horsted Keynes and Seaford operated, was closed on 28 October 1963. Shortly before services were withdrawn 2-BIL unit No 2123 dating from 1938 is seen at Horsted Keynes while LBSCR Terrier 0-6-0 tank No 55 *Stepney* (formerly BR No 32655) prepares to depart for Sheffield Park. One hundred and fifty-two sets of 2-BILs (2-car Bi-Lavatory units, i.e. having a lavatory in each car) were built and the last examples were withdrawn from normal service on 29 July 1971. (*Marcus Eavis/Online Transport Archive*)

An East Grinstead (high level) train via Rowfant and Grange Road prepares to depart from Three Bridges on 2 March 1963. Motive power is provided by SECR class H 0-4-4 tank No 31518 which has clearly attracted two young admirers – perhaps not surprising since, in the author's opinion, these locomotives, as with many of Wainwright's designs, were among the prettiest ever built and enhanced by the distinctive "pagoda" style cab roof and the ornate SECR livery. Sixty-six class H tanks were built at Ashford Works between 1904 and 1915, this particular machine dating from 1908. The sole survivor is No 31263 which, rather appropriately, now works to East Grinstead on the Bluebell Railway. *(Neil Davenport)*

Staying with the former LBSCR, we now move eastwards to Horsham to view an LSWR locomotive type, class M7 0-4-4 tank No 30051, working an auto train to Shoreham-by-Sea via Steyning on 21 May 1956. The Steyning line opened in 1861, followed four years later by the Cranleigh line from Horsham to Guildford. This enabled the LBSCR to poach the LSWR's south coast traffic from Guildford and carry it using its own tracks. However, by the early 1960s little attempt was made to co-ordinate services, resulting in poor connections at Horsham and the closure of both lines, with the Steyning route succumbing on 7 March 1966. Such is progress that anyone wanting to go direct to Shoreham from Guildford now will have to walk! The South Downs Way uses much of the trackbed of both lines. *(Neil Davenport)*

The Guildford - Horsham via Cranleigh line opened on 2 October 1865 but sadly missed its centenary by four months, being closed on 14 June 1965. The last intermediate station southwards was Christ's Hospital (West Horsham), on the Arundel via Pulborough line, built to serve the nearby public school. The picture **above** shows LMS Ivatt class 2 2-6-2 tank No 41301 entering Christ's Hospital a few days before services from Guildford were withdrawn. Arguably the most interesting station on the branch was Baynards, on account of its floral displays and being built to placate the local landowner rather than to serve any other habitation. This view, **right**, of the signal box, surrounded by flower beds in bloom, dates from 10 October 1953. *(Neil Davenport – both)*

Horsham, as well as being reached by the now defunct line from Guildford via Cranleigh, is still served by trains from Three Bridges on the Brighton main line, and also by services via Sutton, Epsom and Dorking. Moving back towards London on the latter line, it is evident from this picture that there were no trains from Horsham to Victoria via Dorking on Sunday 18 September 1955 because the up line between Epsom and Ewell East has disappeared! Wagons filled with ballast in connection with a track re-laying operation occupy the sidings which used to flank the electrified tracks at this leafy spot near Windmill Lane bridge. At the head of the train occupying the northern siding is class N 2-6-0 No 31871 from Bricklayers Arms shed, a locomotive built in 1925 and withdrawn 38 years later. The first vehicle is a ballast plough for levelling. *(Neil Davenport)*

Further up the line through Sutton and Mitcham Junction we come to Streatham where this unusual view was taken on 6 May 1963 and which completes the SR Central section segment. The locomotive is Maunsell class Q 0-6-0 No 30536 built at Eastleigh in October 1938 and withdrawn in January 1964. The Q class of twenty locomotives was Maunsell's final design before retirement and Bulleid was in command by the time the class was constructed. He regarded the type as mediocre and old fashioned and went on to design the more powerful and utilitarian Q1 class in preference to continuing the production of Qs. No 30541 is the sole survivor, thanks to Dai Woodham, and is kept on the Bluebell Railway. *(Paul de Beer/Online Transport Archive)*

It is back to London now for a look at the SR's South Eastern section. Built on a brick viaduct, Elephant & Castle station was opened by the LCDR in 1862 and in this view dating from winter 1957/8, an 8-coach southbound train from Charing Cross or Cannon Street has entered the station. The leading unit, No 4302, was originally built in 1925 as a 3-car set for the Western section's London - Guildford/Dorking services. Soon after the end of the Second World War, when 4-car sets became the norm it was converted into this configuration by the addition of a steel-bodied trailer. *(Julian Thompson/Online Transport Archive)*

On 22 April 1961, the SR cross-divisional 5am freight from Surbiton to Paddock Wood reaches Grove Park headed by SECR class E1 4-4-0 No 31019 piloting class V "Schools" 4-4-0 No 30921 *Shrewsbury*. The E1 was one of 11 Maunsell rebuilds of Wainwright's 26 class Es, this locomotive being rebuilt by Beyer, Peacock & Co of Manchester in 1920. The Schools class 4-4-0s were designed by Maunsell to haul express trains between London - Hastings through the narrow tunnels on that line, forty being built between 1930 and 1935, and three of which have been preserved. They were the last new design of 4-4-0 in Great Britain and the most powerful. *(Charles Firminger)*

Bulleid Q1 0-6-0 No 33002 stands on one of the through lines at Gravesend Central (**above**) ready to form a branch train to Allhallows-on-Sea while an electric unit bound for Charing Cross via Bexleyheath waits in the shadows. Gravesend station was opened by the SER in 1849 and a branch to Port Victoria via Cliffe and Sharnal Street opened in 1882. The SR wrongly anticipated that Allhallows would develop into a popular resort and consequently, in 1932, built a deviation from the Port Victoria line to Allhallows, at a point that became known as Stoke Junction. The Gravesend - Allhallows services were withdrawn on 4 December 1961. The view (**right**) of class H tank No 31519 leaving Cliffe station for Gravesend illustrates the more normal motive power used on the Allhallows shuttle. (*Julian Thompson/Online Transport Archive – both*)

Gillingham-based SECR class C 0-6-0 No 31682 shunts at Farningham Road & Sutton-at-Hone, east of Swanley and on the mainline to Chatham, on 2 November 1959. 109 of these versatile and very successful locomotives were built between 1900 and 1908, this particular example being constructed in 1900 by Neilson, Reid & Co in Glasgow and clocking up a very respectable 61 years of service. One member of the class survives, No 592 (31592), which is active on the Bluebell Railway Farningham Road, opened by the LCDR in 1860, was also the start of a branch to Gravesend West which closed to passengers in 1953 and to freight in 1968. *(Alan Sainty collection)*

Locomotives sandwiched between coaches were relatively uncommon and usually arose when a third or fourth carriage was added to a push and pull two-coach set. Such a configuration is demonstrated in this photograph taken on 22 May 1961 of H class tank No 31177 propelling a push/pull set and hauling a Maunsell open third coach. The train is the 5.23 pm Westerham - Dunton Green service and the location is between Brasted and Chevening Halt. This would now be close to the clockwise carriageway of the M25 since this orbital car park occupies part of the trackbed of the abandoned branch, just west of the junction with the M26, heading towards the romantic-sounding Clackett Lane Services! (*Charles Firminger*)

The 4.5 mile branch to Westerham from Dunton Green, just north of Sevenoaks on the Hastings mainline, was opened by the SER in 1887 with the intention of extending the line a further four miles to Oxted, which never came to fruition. Services were withdrawn on 30 October 1961 despite considerable local opposition and an attempt to lease or buy the line came close to success, but eventually failed. If the scheme had succeeded a section of the M25 would need to have been re-routed! The picture **(left)** shows the "Westerham Flyer" in the care of H class tank No 31322 at Dunton Green and No 31177 **(above)** at the terminus at Westerham. (Alan Sainty collection; Martin Jenkins/OnlineTransport Archive)

A contrast in motive power occurs on the same day in June 1961 at Paddock Wood, the next station east of Tonbridge on the mainline to Dover. Dating back to 1900 when it was built by Sharp, Stewart & Co of Glasgow, class C 0-6-0 No 31716 **(left)** heads a train probably destined for Maidstone West. Built some sixty years later, D6546 **(above)** represents the new order, one of 98 Bo-Bo diesel electrics built by the Birmingham Railway & Carriage Co and known as Cromptons, a reference to their electrical equipment. Later designated Class 33, no fewer than 26 of these capable machines still survive. In the siding behind the station stands a brand new 2-HAP unit awaiting entry into service. *(Marcus Eavis/Online Transport Archive – both)*

In addition to operating services in Kent, the SER also controlled the east/west cross country route at the foot of the North Downs linking Tonbridge, on the SER's main line from London to Hastings, with Redhill, on the LBSCR's Brighton line, and onwards to Guildford (LSWR) and Reading on the GWR's west of England main line. On 19 April 1954, Maunsell U1 class 2-6-0 No 31900 built in 1931 brings a Tonbridge - Redhill train into Penshurst station in Kent. The U1 class totalled 21, with BR numbering in the series 31890-31910, the first locomotive being a rebuild of a River (K class) 2-6-4 tank (see page 83 opposite), followed by twenty new machines. The distinctive slab front above the buffer beam housed the inside cylinder, the U1s being a 3-cylinder development of the 2-cylinder U class. *(Neil Davenport)*

SECR Class U 2-6-0 No 31790 enters North Camp, between Farnborough and Ash, on the Hampshire/Surrey border, hauling the 1.50 pm Reading - Redhill train on 23 February 1963. The locomotive is particularly interesting because it was the prototype River class 2-6-4 tank, named *River Avon* and built at Ashford in 1917. A further twenty such locomotives were built by the SR in 1925-6 but following the Dunton Green (Sevenoaks) disaster of 1927, which was blamed on the instability of these machines at high speed, all were rebuilt as 2-6-0 tender engines in 1928. The pristine platform building has now been replaced by a glorified bus shelter. *(Alan Sainty collection)*

Turning now to the LMR, we start with the former LNWR main line from Euston, nowadays referred to as the West Coast Main Line (WCML). This is Kilburn High Road station on 14 May 1961 with Willesden's Fowler class 4 2-6-4 tank No 42367 hauling empty stock from Euston to Stonebridge Park carriage sidings. One hundred and twenty-five Fowler tanks of this type were built between 1927 and 1934 and the class went on to spawn Stanier and Fairburn versions. No 42367 entered service on 2 September 1929 and was withdrawn on 25 August 1962. Although none has been preserved, there was a very similar looking class built after the Second World War by the LMS at Derby for Northern Ireland, based on the Fowler design rather than the contemporaneous Fairburn engines. One of these locomotives, nicknamed "Jeeps", No 4, which was built in 1947 and withdrawn in 1970, has been saved by the Railway Preservation Society of Ireland. (*Charles Firminger*)

An up freight passes South Kenton station on 27 July 1963 hauled by Watford-based Standard class 2 2-6-0 No 78035. This locomotive was built in 1954 and withdrawn after a mere twelve years of service. Sixty-five were built and were almost identical to the 128 members of the LMS Ivatt-designed class 2 2-6-0s which were still being built when the first of these Standards entered service, making the creation of this new class a somewhat pointless exercise. South Kenton was an additional station opened by the LMS on 3 July 1933 serving the route to Watford Junction operated by both the former LNWR electric services from Euston and the Bakerloo tube line from Elephant & Castle, the latter having reached Watford Junction in 1917. Bakerloo trains ceased to serve South Kenton in 1982 when services were cut back to Stonebridge Park and then re-instated in 1984 following the resumption of services to Harrow-on-the-Hill, but not to Watford Junction (yet!). *(Alan Sainty collection)*

In the post-war period Associated Commercial Vehicles Ltd (ACV) was the holding company for a number of associated bus manufacturing companies including AEC and Park Royal Vehicles, builders of many London buses. British United Traction Ltd (BUT) was an AEC/Leyland joint venture company providing traction equipment for trolleybuses and diesel railcars. Using the expertise of these companies, ACV produced a prototype 3-car set in 1952 powered by two London bus engines which was trialled on various BR and LT branches. The sets offered operational flexibility, being designed to run as three, two or single units, and BR ordered a small fleet. Recently delivered, one such set seems to be attracting the attention of three young loco spotters at Watford Junction on 6 July 1957. (*Marcus Eavis/Online Transport Archive*)

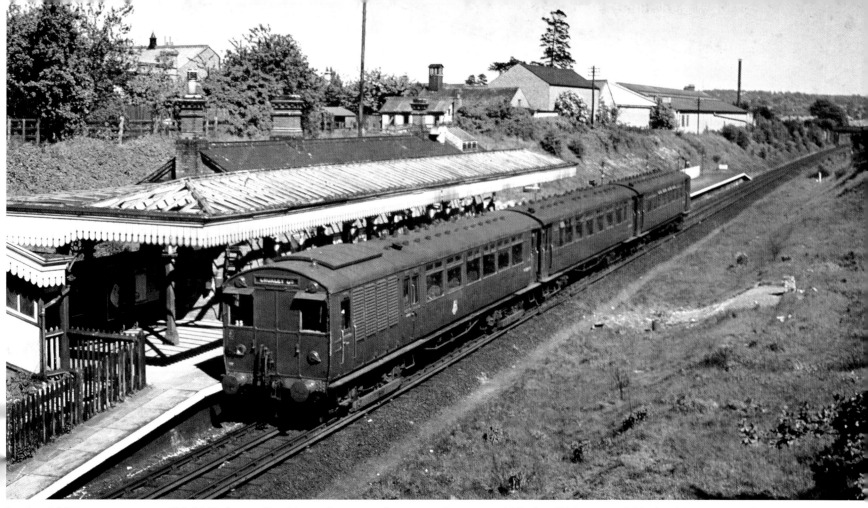

In this 1957 view, a train of LNWR 3-car Oerlikon electric saloon stock stops at Watford West, on a Watford Junction - Croxley Green service. The initial batch of saloon stock was built in 1914 with German electrical equipment sourced from Siemens, but because of the First World War subsequent units built from 1915 onwards had electrical equipment supplied by Oerlikon from neutral Switzerland. The Croxley Green route branched off the former Watford to Rickmansworth line and opened in 1912. Oerlikon stock replaced steam operation when the line was electrified in 1922 and was not itself finally superseded until April 1960, nearly three years after the withdrawal of the rest of this stock. Passenger services on the Croxley Green branch were withdrawn on 22 March 1996 due to a road building scheme and the line was then mothballed pending future re-opening as part of the Croxley Rail Link diverting LT's Metropolitan Line from Watford Met station to Watford Junction (see page 155). (Julian Thompson/Online Transport Archive)

In Great Train Robbery territory a southbound freight has just emerged from the southern end of Linslade tunnel, just north of Leighton Buzzard on 15 July 1961. When the line was widened, two new bores were added on either side of the original double track bore built in 1838, resulting in the unusual situation of the up main line sharing the centre bore with the down slow. The train, which has used the eastern single bore, is hauled by Hughes "Crab" 2-6-0 No 42822. The "Crabs", although they emerged under LMS auspices in 1929, were of Lancashire & Yorkshire Railway (L&Y) appearance, having been designed by their Chief Engineer. Indeed, No 42822, dating from 1929, was one of the 70 from the class of 245 to have been constructed at the former L&Y Works at Horwich. *(Charles Firminger)*

Still at Linslade Tunnel on 15 July 1961, Stanier 8F 2-8-0 No 48549 approaches the centre bore at the head of a northbound pick-up goods. This locomotive was one of 852 such engines constructed by various builders (the big four main line companies and outside contractors) between 1935 and 1946, many being destined for the War Department. Upon nationalisation in 1948, BR found itself with 624 and this number was augmented by various acquisitions from the Military up to 1957, when the total reached 666. The track in the foreground of this and the previous shot is not a siding or branch but the down main line. (*Charles Firminger*)

Backtracking south from Linslade to Berkhampstead, we find an elderly workhorse in the shape of LNWR Bowen-Cooke 7F 0-8-0 No 49106 on a freight train, possibly to do with the WCML electrification scheme, in February 1962. There were various classes of LNWR 0-8-0 of similar appearance and all became generally known as "Super Ds", a reference to their origin and subsequent superheating. The D class were rebuilds of earlier Webb locomotives and further developments of the genre brought about the G1, G2 and G2A classes. No 49106 was built in 1910, converted from class G1 to G2A in 1942 and withdrawn on 15 December 1962. An almost identical G2, No 49395, survives as part of the National Collection. *(Author's collection)*

Standing at Bletchley on 10 May 1958, at the head of a train of suburban stock, is No 10001, Britain's second ever mainline diesel locomotive (the first was its twin, No 10000). Following the end of the Second World War, the LMS planned to use diesels for express services and ordered the aforementioned Ivatt-designed prototypes, with mechanical parts built by the LMS at Derby and electrical parts and power unit provided by English Electric. However, with a mere 1600 hp output, each locomotive was rated only 5P and to haul the heaviest expresses it was necessary to use them in tandem. The LMS managed to launch No 10000 three weeks before Nationalisation but No 10001 did not enter service until July 1948, so always wore BR livery. It was withdrawn in 1966. *(Marcus Eavis/Online Transport Archive)*

Moving across to the former Midland Railway, one of its lesser known local services were those from St Pancras/Kentish Town to Barking. The section from South Tottenham to Woodgrange Park was a joint Midland Railway/LTSR scheme and the image **(above)** depicts passengers on the platform at South Tottenham who may be intending to use the interchange at Barking to travel to the seaside (Westcliffe/Southend). The locomotive is Kentish Town-based Stanier class 3 2-6-2 tank No 40160. South Tottenham was also served by North Woolwich - Palace Gates trains which used a curve (hidden by the train) to reach Seven Sisters. The bridge in the distance carries the Liverpool Street - Enfield line. Woodgrange Park **(right)**, which opened in 1894, was the last station before Barking. The train is headed by another Kentish Town class 3 2-6-2 tank, No 40111. Both views date from 1958 during the line's decrepid days but it is now flourishing. *(Julian Thompson/Online Transport Archive – both)*

There were 733 ex-Government War Department (WD) 2-8-0s in BR stock, built either by North British or Vulcan Foundry. The LNER bought 200 in 1946 and borrowed a further 278, this example, No 90293, being one of the loaned engines. It was built by North British in January 1944 and remained in service until September 1965. In this view, the locomotive is hauling a train of empty coal hoppers near the site of Finchley Road (MR) station, having emerged from Belsize Tunnel. The station, on the main line to St Pancras, opened in 1868 and closed in 1927, although the goods yard remained open until 16 May 1983. A chimney pot and gable end of the station building on Finchley Road is just visible behind the first two wagons. (*Harry Luff/Online Transport Archive*)

Stanier "Jubilee" class 6P5F 4-6-0 No 45614 *Leeward Islands,* which clocked up some thirty years of service from 1934, passes through Mill Hill Broadway station, between Elstree & Borehamwood and Hendon, with an up parcels train. As with Finchley Road, Mill Hill Broadway station opened in 1868 in connection with the MR's London extension to St Pancras. There was an adjacent interchange station, Mill Hill (The Hale), on the GNR's line from Finchley to Edgware. This line was due to be taken over by LT and electrified, but the Second World War intervened and electrification only reached as far as Mill Hill East. The Hale station was closed to passengers on 11 September 1939 as a temporary measure to facilitate the works necessary for it to become part of the LT Northern Line but these works were abandoned and passenger services were not restored. LT bus route 240 provided an adequate replacement service, tickets for which were obtainable from Mill Hill Broadway station well into the 1960s. (*Alan Sainty collection*)

A lengthy up freight train probably bound for Cricklewood trundles along the Midland mainline north of Elstree & Borehamwood in summer 1958 hauled by Crosti-boilered Standard 9F 2-10-0 No 92023, based at Wellingborough. Franco-Crosti boilers, designed by Italian engineers, enabled exhaust gasses, which would otherwise escape from the main chimney, to flow through a feed water heater, which acted as a second boiler, heating the water supply for the main boiler before escaping through a side-mounted chimney located behind the small smoke deflector visible in the picture. The water heater (secondary boiler) was fitted with a small smokebox door immediately below the normal smokebox door. However, these boilers did not produce the efficiencies expected and created maintenance problems, leading to their replacement by conventional boilers. No 92023 was completed in May 1955, rebuilt in September 1961 and withdrawn in November 1967. (*Julian Thompson/Online Transport Archive*)

This St Pancras-bound local train, photographed in 1958, has just left Radlett station and is heading for Elstree & Borehamwood behind Fairburn class 4 2-6-4 tank No 42134, shedded at St Albans. Two hundred and seventy-seven Fairburn tanks were built between 1945 and 1951 and were based on the earlier Stanier tanks, but with a shorter wheelbase and of lighter weight, allowing more flexible operation. With standard gauge steam on BR ending in 1968, inevitably this Derby-built locomotive had a relatively short working life, being completed in January 1950 and withdrawn on 1 April 1967. Two members of the class have been preserved, Nos 42073 and 42085, and these can be found on the Lakeside & Haverthwaite Railway. (*Julian Thompson/Online Transport Archive*)

In order to reach London from the East Midlands, the Midland Railway initially entered into an agreement to use the LNWR's Birmingham to Euston line from Rugby but delays to Midland trains prompted the Company to build a line from Bedford to Hitchin to join the GNR's main line to King's Cross. This opened in 1857 but relations between the two companies deteriorated, forcing the Midland to build its own route from Bedford to London in 1868. Consequently, the Bedford to Hitchin line was relegated from main line to branch status, culminating in the withdrawal of passenger services on 1 January 1962 and freight services from 28 December 1964. On 20 June 1959, Johnson 3F 0-6-0 No 43474, in its 63rd year of service, has arrived at Cambridge Junction, Hitchin, with the 2.33 pm freight from Bedford. (*Charles Firminger*)

The diesels have arrived to replace steam workings on the Bedford - St Pancras services (nicknamed, perhaps derisorily, the "Bedpan Line") in this 1961 shot at Bedford Midland Road (now plain Bedford). The first station was opened by the Midland Railway in 1857 in conjunction with the opening of the Hitchin line and then revamped in 1868 for the Midland's London Extension to St Pancras. This rebuilt station was damaged during the Second World War and was eventually replaced by a new station some 100 yards further north which was opened in 1978. *(Marcus Eavis/Online Transport Archive)*

In addition to the Hitchin branch, another closed line which served Bedford Midland Road was that to Northampton. In a vain attempt to reduce costs and attract more passengers, diesel railbuses were introduced on these lines in 1958. On the Northampton line, the first station after Bedford was Turvey where this photograph of brand new Park Royal four-wheel railbus No M79971 was taken in September 1958. The line opened in 1872 and the last passenger train ran on 3 March 1962. As the name suggests, railbuses had much in common with motor buses and were constructed for BR by a variety of manufacturers such as Bristol/Eastern Coach Works, Wickhams, AC Cars, as well as Park Royal Vehicles. *(Frank Hunt/LRTA)*

We now return to the LNWR to visit Bedford's other station, Bedford St John's, which opened in 1846 as plain Bedford and therefore pre-dates the Midland Road station. St John's, named as such in 1924, was on the so-called "Varsity Line" from Oxford to Cambridge via Verney Junction, which also linked with the Banbury branch through Buckingham and, up to 1936, the LT line to Aylesbury and Baker Street. The complete Varsity line was closed on 1 January 1968 but the section from St John's to Bletchley was retained and still operates, although the station has been resited to enable trains to call at Bedford (formerly Midland Road). The station may however move back to its former site if the Varsity line is reopened, part of which (the Bletchley - Oxford section) is already having preliminary work undertaken on it under the East-West Rail Project. The photograph depicts Bedford St John's in 1959 with coaching stock in the centre road headed by Standard class 2 2-6-2 tank No 84004. *(Marcus Eavis/Online Transport Archive)*

Buckingham station is both quiet and busy in these 1959 pictures. With no footbridge, a passenger crosses the line **(above)** behind BR Derby-built single units Nos 79900 and 79901 coupled together, having arrived with a service from Banbury (Merton Street). These units were meant to save the line from closure, normally operating singly, but in tandem at busier times. On the same day **(right)**, Ivatt class 2 2-6-2 tank No 41275 heads a Bletchley train. Passenger services on the Banbury line ran for the last time on 31 December 1960 and on 5 September 1964 for the Bletchley line, bringing closure to Buckingham station (albeit re-opened for a visit by HM The Queen on 4 April 1966!). The station had opened in 1850, the attractive buildings shown here dating from 1861. Had it not been for objections from the Duke of Buckingham, there would have been a station here, and a railway works, in 1838, on the London to Birmingham main line. Instead, the line was routed through Wolverton. *(Marcus Eavis/Online Transport Archive – both)*

Broad Street station, next door to Liverpool Street station, was the City terminus of the NLR whose original line was between Camden Town and Poplar. The LNWR was a majority shareholder of the NLR and saw this railway as a means of its reaching London Docks. All the railway companies operating around London wanted their own London terminus rather than having agreements to use other railways' terminals, a situation the NLR found itself in, through using Fenchurch Street station. The NLR therefore extended its railway into the City and opened Broad Street station in 1865. At its peak in the late 1890s, the station was the third busiest in London despite handling only suburban services, but competition from electric traction, i.e. trams and the Underground, sent it into a gradual decline, with the last train leaving on 27 June 1986. This photograph of the station was taken on 11 August 1968 and among the road vehicles featured are one Ford Anglia and a Renault Dauphine. (*Marcus Eavis/Online Transport Archive*)

The ex-LNWR Oerlikon saloon stock on the Broad Street - Richmond line was supplemented by LMS Compartment stock of 1926 vintage, as seen here at Acton Central station on 12 October 1957, these units lasting until 1963. The complete line was electrified using the fourth rail system (like the Underground) in 1916, in an attempt by the NLR to compete with electric trams and the Underground, which were taking so much of their business. The current route now uses overhead line equipment, except for the southerly section, and Acton Central is the changeover point for the power supply. Apart from that feature, Acton Central station today is comparatively unchanged from the photograph except for modernisation of the platform lighting. (*Marcus Eavis/Online Transport Archive*)

With the withdrawal of the aged Oerlikon stock in 1957 (except that used on the Croxley Green branch – see page 87), new electric units were introduced on the Broad Street - Richmond line and the Watford Junction DC line. These units were known as Class 501 stock and operated until the service from Richmond was diverted from Broad Street to North Woolwich from 13 May 1985 (it now terminates at Stratford). Kew Gardens station, where this picture was taken in 1959, is also served by LT's District Line and looks much the same today. The station is in a conservation area and the French-designed reinforced concrete footbridge, which dates from 1912, has Grade II listing. (*Marcus Eavis/Online Transport Archive*)

We are now moving across to the former Great Central Railway (GC) which was subsumed within the LNER at Grouping in 1923 and, upon Nationalisation in 1948, into the ER. However, on 1 February 1958, most of the network, including the London end, was transferred to the LMR. This scene outside Marylebone station on 23 August 1966, ten days before ex-GC line services were withdrawn beyond Aylesbury, focuses on a failed locomotive. For historical reasons an ex-LNER locomotive was wanted for the final day of London services beyond Aylesbury on 3 September 1966 and consequently Thompson Class B1 No 61306 (since preserved) was being trialled on the 8.15 am from Nottingham (Victoria) but arrived over one hour late after a poor run. Sadly, it was not considered capable of taking up duty for the GC swansong. *(Bill Piggott)*

The majority of locomotives used on the GC semi-fasts between Marylebone and Nottingham in the latter years were Stanier Black 5 4-6-0s. On 26 August 1966, one such machine, No 44858, hauls the 2.38 pm Marylebone - Nottingham (Victoria) past LT's Neasden station. Opening in 1880, this former Met station ceased to be served by Met trains in 1940, having been transferred to the Bakerloo line in the previous year (and, from 1979, is now part of the Jubilee Line). There was no GC interchange with LT at Neasden (the first being at Harrow-on-the-Hill), except for a few weeks in late 1940 when a temporary platform was erected where the train is passing, due to bomb damage to the track near Marylebone. The lines in the foreground lead to the GC engine shed and to Neasden Junction, linking up with the Cricklewood/Acton Wells cross-region line. *(Bi. Piggott)*

In 1923, the LNER built a single track clockwise loop (circular) line to serve the British Empire Exhibition of 1924/5 with a single station called The Exhibition Station (Wembley) constructed beside the then new (but now replaced) Wembley Stadium. The loop was connected to the Marylebone main line at two junctions near Neasden. Following the closure of the Exhibition the loop continued to be used for special events such as football Cup Finals until 1968, before being abandoned. On 6 May 1961, Fairburn 2-6-4 tank No 42082 approaches Wembley Stadium station (its name since 1928) on Cup Final Day. The result: Spurs 2, Leicester City 0. *(Charles Firminger)*

The LMR, having taken over the ex-GC line in 1958, gradually introduced its own locomotive types in preference to ER locomotives and withdrew the route's express services in 1960, leaving just local and semi-fast trains. In this view Stanier Black 5 4-6-0 No 45238 from Woodford Halse shed stands at Harrow-on-the-Hill on 29 December 1962. The train is running half an hour late on its journey from Nottingham (Victoria) to Marylebone following the first main snowfall of the great 1962/63 winter (which caused the 13-year-old photographer to fall from his bicycle as he returned home over the Hill!). *(Bill Piggott)*

On the penultimate Saturday of the GC, 27 August 1966, "The Dick Flyer" waits at Harrow-on-the-Hill. The train is the 2.38 pm Marylebone - Nottingham hauled by Black 5 No 45267 and the chalked inscription refers to popular GC driver, Dick Hutt, seen leaning out of the cab, who could always be relied upon to give his passengers a good run. By this time, the GC Black 5s were shedded at Colwick, Nottinghamshire, which was transferred from the ER to the LMR on 1 January 1966. This unusual view was obtained by the photographer standing on the saddle of his bicycle to avoid the wire fence. (*Bill Piggott*)

In 1906, a joint agreement came into effect between the GC and the Met over the operation of the Met tracks between Harrow-on-the-Hill and Verney Junction. Stoke Mandeville, the station before Aylesbury travelling northwards, was therefore a joint station until LT services from Baker Street to its subsequent terminus at Aylesbury were cut back to Amersham after 9 September 1961. In the summer of that year, a Woodford Halse - Marylebone train was photographed at Stoke Mandeville hauled by Thompson class B1 4-6-0 No 61120. *(Marcus Eavis/Online Transport Archive)*

Among the ex-LNER classes which used to operate services on the GC were Gresley's powerful class V2 2-6-2s. Nicknamed "Green Arrows" after the name of the doyen of the class (in all eight were named, out of the 184 constructed), these engines were designed for express mixed traffic work, the absence of rear driving wheels providing space for a large firebox and turning these locomotives, arguably, into scaled down Gresley pacifics, which they could nearly match in terms of high speed running. No 60879, dating from 1940, was photographed at Aylesbury on 2 August 1958 on its way to Marylebone. (*Marcus Eavis/Online Transport Archive*)

Framed by a rudimentary footbridge (since replaced by a modern structure) Black 5 No 45289 is ready to depart from Aylesbury on the 4.38 pm Marylebone - Nottingham (Victoria) semi-fast on 13 August 1966. The station was officially named Aylesbury Town to distinguish it from the Aylesbury High Street station on the LNWR branch from the WCML at Cheddington which closed to passengers in 1953 and to freight in 1963. The right hand face of the platform seen here was served by trains on the WR line from Princes Risborough, on which 64-year-old "bubble car" class 121 single unit DMUs currently operate. This line was formerly controlled by the GW & GC Joint Committee while the main line fell under the Met & GC Joint Committee. Consequently, Aylesbury station ended up being managed by a joint committee comprising the two joint committees! *(Marcus Eavis/Online Transport Archive)*

Because of friction with the Met over access to the latter's track prior to the joint agreement, the GC decided to reduce reliance on the Met for access to London by having another joint line, this time with the GWR (the "Alternative Route"). This line ran southwards from Ashendon Junction, located between Bicester and Princes Risborough, and Northolt Junction, being connected to the GC main line at Grendon Underwood Junction to the north and at Neasden to the south. One of the stations on the latter link line was Sudbury & Harrow Road, opened in 1906 and allegedly today's most underused station in London, which is rather surprising in view of its urban location. The station certainly looks tired in this view from summer 1958 of Stanier 2-cylinder 2-6-4 tank No 42588 on a rake of ER coaches. The engine dates from 1936 and worked for 28 years. *(Julian Thompson/Online Transport Archive)*

Neasden-based Standard class 4 2-6-0 No 76037 arrives at Sudbury & Harrow Road on 13 June 1963 with a pick-up goods. Passing a vertically challenged signal, the locomotive is about to shunt coal wagons in the station yard. No 76037 was built in 1954 and, like all the Standard classes, had a short working life, although in this case thirteen years was not a bad innings (by Standard standards!!). It is certainly questionable whether it was worth building the Standards because, even though steam was not predicted to have been eliminated so swiftly, if at all, when the first ones emerged in 1951, it is arguable that all the classes except the unique and highly successful 9Fs, replicated existing types designed by the pre-nationalisation companies. (*Bill Piggott*

The next station west of Sudbury & Harrow Road is Sudbury Hill Harrow which opened in 1906 as South Harrow and was renamed in 1926. Like other stations on this section of the former GC, it had seen better days when photographed here at 3.12 pm on 18 June 1963. The train is a return empty newspaper and parcels van train from Nottingham Victoria to Marylebone hauled by Royal Scot 7P 4-6-0 No 46163 *Civil Service Rifleman*, albeit by this time devoid of its nameplates. The Fowler-designed Royal Scot class was introduced in 1927 and all seventy of the original locomotives were rebuilt by Stanier between 1943 and 1955. There was a 71st Royal Scot which had been rebuilt earlier (in 1935) following its conversion from an experimental high-pressure compound engine. *(Bill Piggott)*

The joint GW/GC line (see page 31) resulted in WR and ER locomotives sharing tracks even in BR days, as illustrated here. **Left**, this local train hauled by Stanier 2-cylinder 2-6-4 tank No 42629 is approaching West Ruislip from Princes Risborough, bound for Marylebone via Neasden. West Ruislip is a terminus of LT's Central Line and the track on the left is the exchange siding for the Tube depot. Ruislip & Ickenham station, as it was originally named, was opened by the GW/GC Joint Railway in 1906 and Central Line services commenced on 21 November 1948. Had it not been for the creation of the Green Belt, the Central Line would have proceeded beyond West Ruislip to Denham. **Above**, in terms of large tank engines, Princes Risborough would normally be frequented by WR 61XX 2-6-2 tanks arriving from Oxford via Thame or from Aylesbury and hence to Maidenhead. However, this is a GC local train, the 11.3 am to Marylebone, headed by Fairburn class 4 2-6-4 tank No 42249 on 5 March 1961. The Fairburn locomotives had a shorter wheelbase than the Stanier tanks from which they were developed and could be distinguished by the gap in the running plate in front of the cylinders. *(Julian Thompson/Online Transport Archive; Charles Firminger)*

Transfer freights from the Midland to the Southern could be seen running alongside the main line between Waterloo and Clapham Junction **(above)** or passing underneath it **(right)**. Cricklewood-based Fowler 3F 0-6-0 "Jinty" tank No 47435, dating from 1926, which clocked up forty years of service, is seen near West London Junction heading for Clapham Junction on 2 April 1961. In the following month Stanier 8F 2-8-0 No 48319, built in 1944 and withdrawn in 1968, brings a freight from Stewarts Lane under the south-west main line heading for Kensington Olympia. West London Junction signal box is on the left of the picture, elevated over the Waterloo tracks, and looking rather peculiar with its metal wartime protective roof. *(Jim Oatway – both)*

Another Cricklewood-based "Jinty" tank, No 47433, brings a van train into South Bermondsey in August 1962. This station is on the old South London line between London Bridge and Queens Road, Peckham, adjacent to Millwall FC's football ground. 422 Fowler-designed Jinties were built between 1924 and 1931, being an LMS development of the MR's 0-6-0 tanks. 417 were acquired by BR in 1948, three having been lost in France during the Second World War and two re-gauged and despatched to Northern Ireland. Ten still exist today, mostly operational or under repair, but No 47433, built by Hunslet in 1926, has not survived. *(Ray DeGroote/Online Transport Archive)*

In addition to transfer freights, there was also inter-regional passenger traffic, as evidenced in this bird's eye view of Stanier Black 5 4-6-0 No 45379 departing from East Croydon with a train from Willesden to Hove on 13 April 1964 and leaving behind a wonderful array of Dinky toys! No 45379 was built by Armstrong Whitworth in Newcastle in 1937 and, remarkably, is still at work on the Southern today carrying the same livery as seen here. Withdrawn from Willesden shed in 1965, it was one of over 200 Barry scrapyard escapees, ending up on the Mid-Hants Railway (Watercress Line), where it has been superbly restored. The photograph was taken from Essex House, SR Central Divisional HQ. *(Charles Firminger)*

Moving across to the ER, we start at King's Cross. The GNR/LNER standard London suburban tank locomotive was Gresley's N2 class. One hundred and seven were built between 1920 and 1929 many of which were fitted with condensing apparatus for use on the Met Widened Lines from King's Cross to Moorgate. They were also used on empty stock workings between King's Cross and Ferme Park carriage sidings in North London, although No 69587 appears simply to be shunting, despite the fact that it is carrying a Hertford destination board on its smokebox door! *(Geoff Morant/Online Transport Archive)*

Gresley N2 0-6-2 No 69584 hauls a down local train near Potters Bar composed of BR Mk1 suburban carriages. This locomotive was built by the LNER in 1929 and was withdrawn thirty years later. The line through Potters Bar and South Mimms (the name of the station when this photograph was taken), was opened by the GNR in 1850, with the terminus at King's Cross being completed two years later. Potters Bar has attracted some railway notoriety in that three serious accidents have occurred there: in 1898 (no deaths), 1946 (2 killed) and 2002 (7 killed). *(Julian Thompson/Online Transport Archive)*

Overlooked by the Hawshead Duty Stone (coal tax obelisk) on top of the embankment at Brookmans Park, Gresley class N2 0-6-2 tank No 69543 heads towards Potters Bar with an up local to King's Cross formed of two Quad-Art sets of articulated carriages. Ninety-eight such sets were built between 1923 and 1929 and the last survivors were in use until 1966. One set, No 74 has been preserved and operates on the North Norfolk Railway. The locomotive, No 69543, was a GNR build from 1921 and provided a creditable forty years of service. One class member survives, No 69523, based on the heritage Great Central Railway. (*Julian Thompson/Online Transport Archive*)

Even Gresley's iconic A4 pacifics could sometimes be found undertaking menial tasks, although the sparkling condition of the locomotive suggests that this may be a running-in turn following overhaul. The date is 28 July 1961 and No 60025 *Falcon*, allocated to King's Cross shed, is working the 7.24 pm King's Cross to Peterborough local near Brookmans Park. Thirty-five A4s were built and six survive in preservation but not No 60025, built in January 1937 and withdrawn in October 1963. King's Cross shed closed on 17 June 1963 following the replacement of the A4s and other pacific classes by diesel locomotives on northbound expresses. *(Charles Firminger)*

Standard 9F 2-10-0 No 92042 trundles through Hitchin in early 1963 with an up coal train from New England Yard, Peterborough to Ferme Park Yard, Hornsey. The freight is just passing Hitchin engine shed which was located alongside the station. In June 1963 No 92042 was transferred to Colwick and it was withdrawn from that shed in December 1965, after less than eleven years active service. However, that was a reasonable life span for this very successful class which simply fell victim to the decision to eradicate steam as quickly as possible after hundreds of new steam engines had been ordered. Only one 9F managed as much as fourteen years service and several worked for as little as five years – a scandalous waste of money. *(Alan Sainty collection)*

Moving across from the Great Northern to the Great Eastern, we start at Liverpool Street. Clocking up a depressingly short working career of a mere fifteen years, smart-looking Thompson B1 mixed traffic 4-6-0 No 61182 is preparing to haul a Cambridge stopping service in this 1958 view. The locomotive was built by Vulcan Foundry and entered service in July 1947 and withdrawn in September 1962. Four hundred and ten B1s were built between 1942 and 1952, 59 of which were named. Two have been preserved, Nos 61264 and 61306, the former being extremely fortunate to have been one of the more unusual residents of the Woodhams care home at Barry. *(Phil Tatt/Online Transport Archive)*

Electric services started on the Liverpool Street to Shenfield line on 26 September 1949, work having commenced on the project and stock ordered before the outbreak of the Second World War, which then delayed completion. Electrification was subsequently extended from Shenfield to Chelmsford and to Southend Victoria in 1956. The 41.5 mile Liverpool Street - Southend route was then converted from 1500V DC to 6.25 kV AC on 4-6 November 1960. The electric units (latterly Class 306) delivered from March 1949 were also converted from DC to AC and the last examples were not withdrawn until 1981. In these views office workers **(left)** wait on Platform 3 to board the arriving electric train at Brentwood station as another stands at Platform 4. At Shenfield **(above)** two units await departure to Liverpool Street. Both pictures were taken on 5 November 1954. *(W C Janssen/Online Transport Archive – both)*

As evidenced on page 130, on 5 November 1954 the photographer was able to ride in the cab of a Shenfield electric unit and took this view of two approaching trains which have just left Liverpool Street: another Shenfield electric and a class N7 0-6-2 tank hauling suburban stock. At the top left of the picture, part of the roof of Granary Junction signal box, which controlled Bishopsgate goods station, can be seen. The elevated freight lines descended to ground level at Bethnal Green. Bishopsgate station opened in 1840 as Shoreditch (it was renamed in 1846) and was the original terminus of the Eastern Counties Railway (later subsumed within the GER) until replaced by Liverpool Street station in 1875, after which it was converted into a goods station. It closed on 5 December 1964 when it burned down. *(W C Janssen/Online Transport Archive)*

Bethnal Green station is at the top of a steep 1 in 70 incline from Liverpool Street, as this picture taken in winter 1957/8 indicates. The train, which is approaching the platform, is a fast line service to Chingford hauled by N7 0-6-2 tank No 69642. This was an LNER-built engine dating from 1926 and withdrawn in 1960. Originally designed by Hill and perpetuated by Gresley, 134 N7s were built between 1915 and 1928, the first twelve emerging under GER auspices and classified L77. One example has been preserved, No 69621, which appropriately was the last locomotive built at Stratford Works. Around the time of this picture being taken some 100 N7s were allocated to Stratford shed and its sub-sheds, but electrification would quickly reduce this huge number. *(Julian Thompson/Online Transport Archive)*

Another fast line service to Chingford waits to depart from Bethnal Green in winter 1957/8, hauled by N7 0-6-2 No 69624. This was also an LNER-built locomotive, originating in 1925 and lasting in service until December 1958. Bethnal Green was the location of the Second World War's worst civilian disaster on 3 March 1943 when 173 people were crushed to death entering the newly built Tube station which at that time had no tracks laid and was being used as an air raid shelter. Ironically, there was no bombing at the time, just an air raid siren test sounding, but the local population, mainly women and children rushed for shelter. Someone tripped, causing a catastrophic domino effect. It took fifty years for a permanent memorial to be erected. *(Julian Thompson/Online Transport Archive)*

Class N7 No 69653 heads a Chingford train comprising two Quint-Art sets between Hoe Street and Wood Street in 1958. The posts for the overhead electric system have just been fixed in position in preparation for the AC electrification of the lines from Liverpool Street to Chingford, Enfield Town, Hertford East and Bishops Stortford with effect from 21 November 1960. The GER had first contemplated electrification at the end of the First World War but abandoned the proposal in light of the projected costs, opting instead for an extraordinarily intensive steam service, achieved through such means as new signalling and trackwork as well as alterations at termini to facilitate locomotive servicing and improve passenger flows. (Julian Thompson/Online Transport Archive)

This 1958 view of the approach to Chingford station luckily just precedes the installation of the overhead electric system which has now turned this open vista into a spider's web of wires and supports. Even the GER signal box which dated from 1920 has been demolished. However, the Victorian station building survives. Chingford was designed as a through station with the intention of the line continuing to High Beech but Queen Victoria reputedly vetoed this proposal on the grounds that it would spoil her beloved Epping Forest. N7 No 69646 is setting off for Liverpool Street with a single Quint-Art set. *(Julian Thompson/Online Transport Archive)*

Class N7 No 69626 hauls a down Liverpool Street - Enfield train composed of a Quint-Art set near Stoke Newington in summer 1958. The upright supports for the overhead electrification have just been put into position. The intensive steam-operated services referred to on page 135, introduced in 1920, were nicknamed "Jazz services" in reference to the yellow (for first class) and blue (for second class) painted panels above the carriage windows, intended to speed up passenger boarding. Locomotive No 69626 was built in 1925 and withdrawn in 1959. *(Julian Thompson/Online Transport Archive)*

Enfield Town was reached in 1849 by a now abandoned line from Edmonton (subsequently renamed Angel Road) via a new Edmonton station (later named Lower Edmonton Low Level) until a shorter route via Seven Sisters and a new high level station (later called Edmonton Green) was opened in 1872. Seven Sisters also served the branch to Palace Gates (Wood Green) and the line to Cheshunt, Broxbourne and Cambridge. With the electrification paraphernalia now in place ready for the changeover from steam to electric trains in November 1960, class N7 No 69668 **(above)** brings an Enfield - Liverpool Street train into Seven Sisters and class-mate No 69670 **(right)** prepares to take a Liverpool Street service out of Enfield Town. *(Marcus Eavis/Online Transport Archive – both)*

Walthamstow Marshes are enhanced by the passage of LNER class B12 4-6-0 No 61572 at Copper Mill Junction. Based at Norwich, the locomotive is hauling the 7.15 pm return excursion from Liverpool Street to Swaffham on 16 May 1961. No 61572 was built by Beyer, Peacock & Co for the LNER in 1928, one of a batch of ten following on from 71 (one replaced an accident victim) Holden GER class S69s constructed between 1911 and 1921. The Claud Hamilton class of 4-4-0s were struggling with the increasingly heavy trains on the GER, hence the need for a more powerful 4-6-0 type. The S69s/B12s were still relatively small but were well suited to GER lines with their low axle loading and short wheelbase. Latterly, No 61572 was the Norwich shedmaster's "pet" who kept it in revenue earning service long enough for it to be purchased for preservation. The locomotive was withdrawn on 20 September 1961, having outlived the rest of the class by almost two years, and is currently based on the North Norfolk Railway. *(Charles Firminger)*

Photographed from the footplate of a passing locomotive, class B1 4-6-0 No 61311 is standing in front of a cooling tower at Angel Road gas works, Edmonton, having brought a delivery of coal from Victoria Docks to this local coal gas plant which operated until 1972. Though externally in reasonable condition, the locomotive was deemed unfit for passenger work at this time due to poor springing. Angel Road is on the line to Hertford East/Bishop's Stortford via Tottenham Hale which was electrified from 5 May 1969. *(Gordon Wells)*

The LNER class J15 (formerly GER class Y14) 0-6-0s were built between 1883 and 1913 and, given that they numbered no fewer than 289, we are particularly fortunate to have here a photograph of the sole survivor, LNER No 7564/BR No 65462, which is based on the North Norfolk Railway. Built in 1912 at Stratford Works, this locomotive spent most of its life in East Anglia but moved to Stratford shed in January 1961, from where it was withdrawn on 16 September 1962. The locomotive and its train of brake vans are occupying the down road at Ponders End, near Enfield, which is the following station northwards after Angel Road. *(Bill Timmings/Gordon Wells collection)*

The powerful Thompson class L1 2-6-4 tanks were designed for heavy suburban passenger work, with the pioneer, LNER No 9000, being built in 1945 but the remaining 99 emerging under BR auspices between 1948 and 1950. Surprisingly, BR numbered them 67701-67800, rather than 67700-67799. The L1s were not much loved, experiencing a number of problems over their relatively short lifespan (all being withdrawn by 1962) including leaking tanks and axlebox overheating. Several were allocated to GE lines but, arguably, much of their work could just as easily have been undertaken by class N7 0-6-2s. In this view taken in 1958 when electrification infrastructure was starting to appear, an unidentified L1 is approaching Ware station with a Liverpool Street - Hertford East service. The brick building on the left is the Wickham Works (think trolleys and railbuses!)
(Julian Thompson/Online Transport Archive)

An unidentified class L1 2-6-4 tank has just left Broxbourne station and is approaching Broxbourne Junction with a down train from Liverpool Street, probably bound for Bishop's Stortford, in 1958. Broxbourne Junction was where the branch to Hertford East, which opened in 1843, diverged from the GE mainline to Bishop's Stortford and Cambridge. The Stratford to Broxbourne section opened in 1840 and the railway reached Bishop's Stortford in 1845. The line between Broxbourne and Bishop's Stortford was electrified in 1968. *(Julian Thompson/Online Transport Archive)*

LNER Sandringham class B17 No 61653 *Huddersfield Town* enters Bishop's Stortford station on 4 July 1959 with an up Cambridge - Liverpool Street train. The three-cylinder B17s were introduced in 1928 and were constructed up to 1937, totalling 73 machines in all. No 61653 was built in 1936 and fitted with a B1-type boiler in 1954. Most of the class were named after stately homes or football clubs, engines featuring the latter having a small football (just visible here) beneath the Club's name. All were withdrawn by 1960 and none preserved, but a start has been made on building a new one. (*Charles Firminger*)

Photographed by the fireman, Standard Class 4 2-6-4 tank No 80071 is working an afternoon diagram which started with a train from Stratford Low Level to North Woolwich where this picture was taken. The service is about to depart for Palace Gates (Wood Green), leaving behind the previous train engine, an N7 tank. No 80071, which has suffered a bent running plate, was built at Brighton Works in 1953 and had been transferred to Stratford after being displaced from the LTSR lines following electrification. The Stratford - Palace Gates line was closed to passengers on 7 January 1963 and the line to North Woolwich on 10 December 2006. The station opened in 1847 and the present terminus building dating from 1854 was latterly a railway museum which closed in 2008. *(Gordon Wells)*

The normal motive power for North Woolwich to Palace Gates trains in the latter days of the service were N7 class 0-6-2s or L1 2-6-4 tanks. Here we see a Palace Gates-bound N7 hauling a Gresley Quint-Art set passing the south end of the huge Temple Mills marshalling yard, between Stratford and Leyton, which in the 1950s was handling around 4,500 wagons per day. This view is from Ruckhold Road bridge looking south-east soon after the tracks had been reconfigured and following the opening, on 29 June 1958, of Manor Yard signal box, seen on the left. Around this time, Temple Mills was described as the nation's most modern marshalling yard utilising the latest electronic devices for the efficient handling of freight. *(Julian Thompson/Online Transport Archive)*

The alternative to the Liverpool Street - Southend (Victoria) railway via Shenfield is the former LTSR line from Fenchurch Street to Southend (Central). This route originally reached Southend only via Tilbury but a second, more direct route via Upminster, was completed in 1888. The LTSR ceased to be an independent company in 1912 when it was acquired by the Midland Railway which in turn became part of the LMS in 1923. In 1949 it became part of BR's Eastern Region. Steam gave way to electric units in June 1962 when the last of the 37 Stanier 3-cylinder 2-6-4 tanks (Nos 42500-36) built in 1934 exclusively for the LTSR line, was withdrawn. The need for powerful locomotives on these commuter services is demonstrated by this view **(left)** of Shoeburyness-based 2-6-4 No 42509 approaching Westcliff with an eleven-coach suburban train. Excursion traffic to Southend also used this line, as evidenced **(above)** by Fairburn 2-6-4 tank No 42254 on an up train soon after leaving Westcliff station. In both views, concrete bases for electrification poles can be seen. *(Julian Thompson/Online Transport Archive – both)*

We conclude our survey of the ER with transfer freights hauled by two consecutively numbered, Hornsey-based Gresley J50 0-6-0 tanks dating from 1927. Some cross-regional transfer freights from the Eastern Region to the Southern would travel from Ferme Park sidings via the Met's Widened Lines from King's Cross to Farringdon and then use Snow Hill tunnel beneath Smithfield Market, emerging just north of Blackfriars station. **(above)** On 20 June 1959, No 68970 is waiting near Holloway to take a freight over the Widened Lines while, a year or so earlier, No 68971 **(right)** is bringing a freight through Wandsworth Road, possibly from Hither Green and destined for Ferme Park via the Widened Lines.. (Charles Firminger; Harry Luff/Online Transport Archive)

For the final segment of this local train review, we take a look at some LT operations, starting at the outer reaches of its empire. The Met line from Aylesbury to Baker Street/Aldgate (cut back from Verney Junction and Quainton Road in 1936) was originally intended to be part of a through route linking Manchester with London and France (via a Channel Tunnel). Until September 1961, when LT electrified the Rickmansworth - Amersham section and withdrew from Aylesbury, LT's electric locomotives were replaced by steam over the final section as shown in these pictures. On 5 March 1961 Fairburn class 4 2-6-4 No 42090 **(left)** prepares to leave Aylesbury with the 10.30 am departure for Baker Street on 5 March 1961 and on a separate occasion another Fairburn tank, No 42222 **(above)** calls at Chorley Wood with a London-bound train. *(Charles Firminger; Julian Thompson/Online Transport Archive)*

The branch from Chalfont & Latimer (called Chalfont Road until 1915) to Chesham was conceived as a through line to the LNWR at Tring when it opened in 1889 and only became a branch in 1894 when the Met reached the Aylesbury & Buckingham Railway's station at Aylesbury and proposals for an extension were aborted. Although there were a few through trains from Chesham to London, most services consisted of a shuttle between the two stations, worked as push and pulls from 1940 using one of two three-car units of Ashbury stock (four of these cars now operating on the Bluebell Railway). From 1937, the LNER, and eventually the LMR, provided the motive power until the last steam service operated on 12 September 1960 followed by electrification. The conductor rails are already in place in this view of Ivatt 2-6-2 tank No 41284 approaching Chalfont & Latimer. The shuttle ceased operation on 11 December 2010 with the introduction of fixed 8-car S8 stock which could not be accommodated in Chalfont's bay platform, so only through services to London now operate on the branch. (Julian Thompson/Online Transport Archive)

Another Met branch which was due to be extended, albeit only about one mile, was that to Watford which leaves the Met's Rickmansworth/Amersham line just north of Moor Park. The Met formed a joint committee with the LNER to build the Watford branch which opened on 4 November 1925, the Met providing electric trains and the LNER steam trains. The local council would not allow the railway, on its existing course, to continue into Watford town centre so Watford Met had to be inconveniently sited one mile away. Undeterred, the Met purchased 44 Watford High Street for their proposed Watford Central terminus but there is no station there – but you can visit it for a drink because the building is now the *Moon under Water* public house! Ironically, the Met line is at last due to be extended, using part of the mothballed Croxley Green branch (see page 87), to reach Watford Junction, whereupon Watford Met station will close. This photograph depicts two Baker Street trains composed of T stock units waiting for passengers. *(Julian Thompson/Online Transport Archive)*

Moor Park station, where these two pictures were taken, stands between Northwood and Rickmansworth. Opening as Sandy Lodge on 9 May 1910 and named after the nearby golf course in an attempt to attract the middle classes to "Metroland", Moor Park station had to be reconstructed when tracks between Harrow North Junction and Watford South Junction (just north of Moor Park) were quadrupled in 1959-61. Until LT ceased to serve Aylesbury in September 1961, trains between there and Baker Street/Aldgate were composed of "Dreadnought" compartment stock designed in 1910 and hauled southwards from Rickmansworth by Bo-Bo electric locomotives dating from 1922-23. In 1958, No 2 *Thomas Lord* (previously named *Oliver Cromwell*) is seen **(above)** approaching Moor Park on an Aylesbury - Baker Street working and **(right)** in May 1955 No 6 *William Penn*, still in wartime grey livery, is entering Moor Park on a down train. *(Julian Thompson/Online Transport Archive; Ray DeGroote/Online Transport Archive)*

Northwood Hills, the station between Pinner and Northwood, was opened on 13 November 1933 and this picture provides another opportunity to view Met Bo-Bo electric locomotive No 6 *William Penn* as it enters Northwood Hills with a Baker Street service from Aylesbury. By 1957, when this photograph was taken, the locomotive had lost its wartime grey livery which had been introduced in 1943, and regained maroon livery. The original ornate bronze nameplates had been melted down for the war effort but replacement ones were cast when maroon livery was restored from October 1953. Twenty of these machines were built, two of which survive: No 5 *John Hampden* on static display at the LT Museum at Covent Garden and No 12 *Sarah Siddons*, retained in working order. *(Julian Thompson/Online Transport Archive)*

You have seen Harrow-on-the-Hill station before, on pages 110 and 111, in the Great Central segment but now it is the turn of the Met, with a Watford-bound train arriving, formed of T stock dating from 1927-30. With their wooden mouldings, these carriages contrast with the smooth steel-panelled bodies of the final batch from 1932-3, as illustrated on page 155. The station was opened as plain Harrow in 1880 following the Met's extension from Willesden Green and received its current name in 1894 as a marketing ploy, since it is not located in that part of Harrow (the old town) which is on the hill! Electrification reached Harrow, together with the Uxbridge branch which bifurcated just north of Harrow, in 1905. Harrow became the electric-to-steam changeover point until electrification was extended to Rickmansworth in 1925. *(Ray DeGroote/Online Transport Archive)*

The Met's branch from Harrow-on-the Hill to Uxbridge became operational on 4 July 1904, with electric services starting on 1 January 1905. In 1910, the District opened up an extension from South Harrow to Rayners Lane which enabled it to run services to Uxbridge as well (these becoming Piccadilly Line trains from 1933). Eastcote station, seen here, opened on 26 May 1906 in response to local housing development and was rebuilt to a Charles Holden design in 1939 (his familiar cube-shaped brick and glass ticket hall can be seen on the left). From 1950, Met trains to Uxbridge consisted of the distinctive F stock dating from 1920-1 which, with its rapid speed and acceleration, was particularly suitable for the non-stop sections to Baker Street. This shot of F stock was taken in 1959. *(Julian Thompson/Online Transport Archive)*

A southbound Northern Line train composed of 1938 tube stock calls at West Finchley, on the High Barnet branch, on 13 October 1957. Apart from the actual train, this scene is virtually unchanged today. The line was opened by the GNR on 1 April 1872 with two intermediate stations. West Finchley was added on 1 March 1933 by the LNER with no expense spared (!): just recycled materials, e.g. the footbridge, from closed stations and very basic wooden buildings. Northern Line electric services over the branch, integrated with the remainder of the network, commenced on 14 April 1940 and the LNER discontinued its steam service on 2 March 1941. LT withdrew the last of its 1938 stock from normal service in 1988, these trains achieving a remarkable fifty years of service on the Northern Line. Even more amazing is that some units are still in service today on the Isle of Wight, making them the oldest passenger stock operating timetabled services on Network Rail. *(Marcus Eavis/Online Transport Archive)*

A District Line service from Upminster to Wimbledon calls at Plaistow in May 1955. The train is composed of Q stock created in 1938 and consisting of vehicles of various styles and age. The unit at the front of the train is one of 50 G stock motor cars built in 1923. Following replacement of their passenger-operated sliding doors, 48 of them were re-designated as Q23 units (the missing two were adapted to run on the Acton Town - South Acton branch). The second carriage in the depicted train is a flared-sided vehicle built in 1938. Plaistow station was opened by the LTSR on 31 March 1858 on its new direct route from Fenchurch Street to Southend avoiding Stratford and Forest Gate. The District Railway reached Plaistow on 2 June 1902. On the right of the picture is the former LTSR line, trains ceasing to call at Plaistow when steam was replaced by electric in 1962, at which point Plaistow steam shed was also closed. That side of the platform is now fenced off since the station is only served by District Line trains. *(Ray DeGroote/Online Transport Archive)*

Two LT Central Line trains composed of Standard Stock are viewed at Stratford from the cab of a Shenfield electric bound for Liverpool Street on 5 November 1954. 1,460 units of Standard Stock were built from 1923 to 1934 and were the first tube trains to have air-operated sliding doors. This stock was retired from the Central Line in 1963 but 43 were sold to BR in 1966-7 for use on the Isle of Wight, lasting in service until 1989, whereupon they were replaced by 1938 stock. Stratford station opened for Central Line trains on 4 December 1946 following the opening of the 4.5 mile extension from Liverpool Street. The Line was extended beyond Stratford to Leyton in 1947, reaching Epping in 1949, which at that time was the limit of electrification over what were mainly ex-GER tracks. With the western extension to West Ruislip, the Central Line became the longest Tube line covering 46 miles and serving 49 stations. *(W C Janssen/Online Transport Archive)*

We end this album with three views of a Central Line outpost, the Epping - Ongar line. This was opened by the GER on 24 April 1865 and transferred from BR's Eastern Region to LT following completion of electrification from Loughton to Epping. LT services commenced on 25 September 1949 but ended on 30 September 1994 and the line has now become a heritage steam railway. From 1949, LT hired a steam auto train from BR using Holden F5 2-4-2 tanks and this arrangement lasted until 16 November 1957, after which an electric shuttle train operated following electrification to Ongar. However, through trains composed of eight cars were unable to operate because of insufficient voltage, so a 4-car shuttle service, without the benefit of steam heating and requiring a change of trains at Epping, was the best that could be offered. The auto train is encapsulated **(above)** in this summer 1957 view of the service steaming out of Ongar over the seven-arch viaduct across the Cripsey Brook hauled by F5 No 67200, a scene which will one day be re-enacted following the completion of a replica F5 locomotive currently under construction. For the first few years of electrification LT Standard Stock was used and this summer 1962 shot **(right)** depicts one such train entering North Weald station on its way to Epping, with the masts of Ongar Radio Transmitter Station prominent in the background. *(Julian Thompson/Online Transport Archive; Marcus Eavis/Online Transport Archive)*

A fitting end to the book is provided by the rear of one of the Epping-Ongar push and pull sets leaving North Weald station in May 1955, viewed from a second set, both hauled by F5 2-4-2 tanks. North Weald was one of two intermediate stations on the branch, the other being Blake Hall, but in the light of service reductions, the passing loop at North Weald became redundant and was removed in 1976. The heritage Epping Ongar Railway have now reinstated the loop to enable two trains to operate on the line. The former GER signal box dating from 1888, part of which is visible on the left of the picture, has been fully restored, including the installation of an original Saxby & Farmer 21-lever frame. *(Ray DeGroote/Online Transport Archive)*

# INDEX OF LOCATIONS
(NB. Some locations are not exact)

Acton Central, 105
Angel Road, 141
Ashurst, 8
Aylesbury Town, 113, 114, 152
Barking, 1
Battersea Park, 57
Baynards, 69
Beaconsfield, 29, 30
Bedford Midland Road, 99
Bedford St Johns, 101
Berkhampstead, 90
Bethnal Green, 132, 133, 134
Bishops Stortford, 145
Bletchley, 91
Brasted, 77
Brentwood, 130
Broad Street, 104
Brookman's Park, 126, 127
Broxbourne, 144
Buckingham, 102, 103
Caledonian Road & Barnsbury, rear cover
Chalfont & Latimer, 154
Chingford, 136
Christ's Hospital, 4, 66
Holloway, 150
Horsham, 67
Horsted Keynes, 64
Hurst Green Junction, 51
Kew Gardens, 106
Kilburn High Road, 84
Kings Cross, 124
Linslade, 88, 89
Liverpool Street, 129
London Bridge, 54, 55, 56
Maidenhead, 24
Marylebone, 107
Mill Hill Broadway, 95
Moor Park, 156, 157
Neasden, 108
North Acton, 26
North Camp, 83
Northolt, 28
North Weald, 165, 166
Northwood Hills, 158
North Woolwich, 146
Old Oak Common, 12
Ongar, 164
Oxted, 50
Chorley Wood, 153
Clapham Junction, 42, 43, 44, 45, 46, 47
Cliffe, 75
Cowley, 20
Cricklewood, 38
Drayton Green, 14
Dunton Green, 78
Eastcote, 160
East Croydon, 58, 59, 123
East Grinstead, 64
Elephant & Castle, 72
Elstree, 96
Enfield Town, 139
Epsom, 70
Farningham Road, 76
Finchley Road, 94
Gravesend, 74
Greenford, 27
Grove Park, 73
Guildford, 53
Hanwell, 15
Harrow-on-the-Hill, 110, 111, 159
High Wycombe, 30, 31
Hitchin, 98, 128

167

Paddington, 10
Paddock Wood, 80, 81
Penshurst, 82
Plaistow, 162
Ponders End, 142
Potters Bar, 125
Princes Risborough, 34, 35, 119
Radlett, 97
Reading, 25
Richmond, 49
Royal Oak, 11
Seven Sisters, 138
Shenfield, 131
Slough, 22, 23
South Acton, 39
South Bermondsey, 122
Southall, 16, 17, 18
South Kenton, 85

South Tottenham, 92
Stoke Mandeville, 112
Stoke Newington, 137
Stratford, 163
Streatham, 71
Sudbury & Harrow Rd, 115, 116
Sudbury Hill, Harrow, 117
Surbiton, 50
Temple Mills, 147
Three Bridges, 66
Tunbridge Wells West, 62, 63
Turvey, 100
Uxbridge Vine Street, 21
Walthamstow, 140
Wandsworth Road, 151
Ware, 143
Waterloo, 36, 37
Watford Junction, 86

Watford (Met), 155
Watford West, 87
Wembley (Exhibition), 109
West Byfleet, 52
Westcliff-on-Sea, 148, 149
West Drayton, front cover, 19
West Ealing, 13
Westerham, 79
West Finchley, 161
West London Jct, 40, 41, 120, 121
West Ruislip, 118
Weybridge, 51
Wimbledon, 48
Wooburn Green, 32, 33
Woodgrange Park, 93
Wood Street, 135